OXFORD EARLY CHRISTIAN STUDIES

General Editors

THE OXFORD EARLY CHRISTIAN STUDIES series includes scholarly volumes on the thought and history of the early Christian centuries. Covering a wide range of Greek, Latin, and Oriental sources, the books are of interest to theologians, ancient historians, and specialists in the classical and Jewish worlds.

Titles in the series include:

The Asketikon of St Basil the Great
Anna M. Silvas (2005)

Marius Victorinus' Commentary on Galatians
Stephen Andrew Cooper (2005)

Asceticism and Christological Controversy in Fifth-Century Palestine
The Career of Peter the Iberian
Cornelia B. Horn (2006)

Marcellus of Ancyra and the Lost Years of the Arian Controversy 325–345
Sara Parvis (2006)

The Irrational Augustine
Catherine Conybeare (2006)

Clement of Alexandria and the Beginnings of Christian Apophaticism
Henny Fiskå Hägg (2006)

The Christology of Theodoret of Cyrus
Antiochene Christology from the Council of Ephesus (431) to the Council of Chalcedon (451)
Paul B. Clayton, Jr. (2006)

Ethnicity and Argument in Eusebius' *Praeparatio Evangelica*
Aaron P. Johnson (2006)

Union and Distinction in the Thought of St Maximus the Confessor
Melchisedec Törönen (2007)

Contextualizing Cassian
Aristocrats, Asceticism, and Reformation in Fifth-Century Gaul
Richard J. Goodrich (2007)

Ambrosiaster's Political Theology
Sophie Lunn-Rockliffe (2007)

Coptic Christology in Practice
Incarnation and Divine Participation in Late Antique and Medieval Egypt
Stephen Davis (2008)

Possidius of Calama
A Study of the North African Episcopate in the Age of Augustine
Erica T. Hermanowicz (2008)

Justinian and the Making of the Syrian Orthodox Church
Volker L. Menze (2008)

The Christocentric Cosmology of St Maximus the Confessor

TORSTEIN THEODOR TOLLEFSEN

OXFORD
UNIVERSITY PRESS

This book has been printed digitally and produced in a standard specification in order to ensure its continuing availability

OXFORD
UNIVERSITY PRESS

Great Clarendon Street, Oxford OX2 6DP
United Kingdom

Oxford University Press is a department of the University of Oxford.
It furthers the University's objective of excellence in research, scholarship, and education by publishing worldwide.
Oxford is a registered trade mark of Oxford University Press in the UK and in certain other countries

Reprinted 2012

British Library Cataloguing in Publication Data
Data available

Library of Congress Cataloging in Publication Data
Data available

ISBN 978-0-19-923714-2

For my children,
Nikolai and Juliane

Preface

For a start I should like to clear up a possible confusion regarding the title of this book. In 2000 I defended my thesis *The Christocentric Cosmology of St Maximus the Confessor—a study of his metaphysical principles* (Acta humaniora 72, University of Oslo, Unipub forlag, 2000) for the degree of *doctor philosophiae* at the Faculty of Arts, the University of Oslo (Norway). At the time of the defence I received a number of copies of the printed version of the thesis, some of which were spread abroad, in Europe and North America. When a revised version was planned to be published by Oxford University Press, I did not want to change the main title, because the term 'Christocentric cosmology' is my own invention and it describes well the contents of my book. However, some Maximus scholars refer to the original thesis in books and articles published in recent years, so that the reader should be aware that all references to Tollefsen, *Christocentric Cosmology* before the publication of this book by the Oxford Early Christian Studies, is to the thesis.

The title indicates the contents of this book. It is an investigation into the structure of St Maximus' cosmology or metaphysical conception of the cosmos as centred in Christ. When I started my work I had the great luck to discover that not very far away, just outside of Stockholm (Sweden), the well-known Maximus scholar Lars Thunberg lived in retirement. For some years I enjoyed the cooperation and friendship of Lars, and I even had the pleasure of being guest in his and his wife Anne-Marie's home in Sigtuna. At first Lars was sceptical about the term Christocentric cosmology, but eventually he came to appreciate it, and even commented on my idea of 'Christocentricity' with approval in the revised Swedish translation of his *Man and the Cosmos* (St Vladimir's Seminary Press: New York, 1985), viz. *Människan och kosmos* (Artos bokförlag, 1999). I am grateful to Lars because he introduced me to the world of Maximus scholarship, and even though an established scholar he was a dynamic thinker, never afraid to change his own views if he found good reasons to do so. I should also like to thank my friends and colleagues in the

Medieval Seminar at the Department of Philosophy in Oslo, together with whom I have spent a great deal of time discussing Greek Church Fathers, St Maximus in particular. I am grateful to Fr Andrew Louth for encouraging me to send the manuscript to Oxford University Press, and for the generous evaluation of my original work. I also would like to thank another Maximus scholar of the first rate, Paul Blowers, for his kind support in this process.

I dedicate this book to my children, Juliane and Nikolai, a constant source of joy and happiness to me.

T. T. T.

Contents

List of Abbreviations

CAG	*Commentaria in Aristotelem Graeca*, Berlin 1882–1909.
Capita 150	Saint Gregory Palamas: *The One Hundred and Fifty Chapters*, edited and translated by R. E. Sinkewicz, Toronto 1988.
CCSG	*Corpus Christianorum, Series Graeca*, Brepols, Turnhout.
DN	Dionysius the Areopagite: *De Divinis nominibus*.
Enn.	Plotinus, *Enneads*.
NPNF	*Select Library of the Nicene and Post-Nicene Fathers of the Christian Church*, originally published in 1886, reprinted by Hendrickson Publishers, Peabody, Mass. 1995 (second printing).
PG	*Patrologiae cursus completus, series graeca*, ed. J. P. Migne, Paris 1857–66.
SCh.	*Sources chrétiennes*, Paris 1940–.
ST	Thomas Aquinas: *Summa theologiae*.

The Works of St Maximus

Ad Thal.	*Quaestiones ad Thalassium*
Amb.	*Ambiguorum liber de variis difficilibus locis Sanctorum Dionysii Areopagitae et Gregorii Theologii*
Cap. gnost.	*Capita theologica et oeconomica*
De char.	*Centuriae de charitate*
LA	*Liber asceticus*
Myst.	*Mystagogia*
Or. Dom.	*Orationis Dominicae expositio*
Pyrrh.	*Disputatio cum Pyrrho*
Qu. dub.	*Quaestiones et dubia*
Th. pol.	*Opuscula theologica et polemica*

1

Introduction

The term Christocentric cosmology adequately describes the world-view of St Maximus the Confessor (580–662). He thinks of the whole natural cosmos as made because of a Trinitarian motif, by the Son of God, with Him as the centre of all created being, and with a view to the establishment of communion between created and uncreated being in Christ, the Logos.

If such is the case, one might still wonder what is the point of writing a book on it. Maximus' 'system' represents an impressive intellectual effort. It contains a lot of ideas prepared in his predecessors, but worked out by him as his philosophy (see § II below). Like the philosophy of Plato and Aristotle, Plotinus and Augustine, it belongs, for sure, to the past. But like their philosophies it has virtues pointing beyond the ancient world, even into the contemporary world. Maximus' thinking of God and creation, of creation as centred in Christ, and of a connection between cosmology and soteriology makes interesting sense of Trinitarian theology and of God's love and perfection. It also makes a strong motivation for an ecological consciousness as one of the major ethical challenges confronting modern human beings. In Maximus' system one finds God's remoteness from and closeness to the created cosmos defined philosophically, but the meaning of this remote and close God in Maximian theology is settled in the mystery of love. If God is God, the divine must be honoured in its majestic remoteness beyond all creatures. If God is God, according to Maximus, God's majestic beyond is all the more to be honoured because in the Divinity's perfect inner life there occurs eternally a motif for creative and salvific action: *philanthropia*, love of human beings.

All of God's activity has one single purpose, viz. to unite the world to Himself. Maximus asserts that the creation and ordering of the world is an embodiment (ἐνσωμάτωσις) of the Logos:[1] 'Always and in all God's Logos and God wills to effect the mystery of His own embodiment.' The Logos is embodied in the world by certain *logoi* that come from Him. These *logoi* of beings are a kind of divine Ideas which, taken together, constitute the divine plan for the created cosmos. On the basis of this plan, as it is actualized in a world consisting of intelligible and sensible beings, the foundation is laid for a cosmic conversion (ἐπιστροφή) to God. The *logoi* belong to the Logos and this Logos/*logoi*-conception is, then, the backbone of Maximus' world-view.

In the following chapters I will try to develop Maximus' thought in a systematic way, analytically and synthetically. That is, I shall present the 'system' as it can be extracted ('synthesized') from his works, without taking into consideration any chronological development. I shall further ask analytically what is the relevance of Maximus' principles and whether his arguments are sound within the context they are put forward.

After this introductory chapter (Chapter 1) I shall make some comments on the philosophical doctrine of exemplarism and Maximus' doctrine of creation (Chapter 2). Then follows his theory of the Logos and the *logoi* as the central theological (Trinitarian) principles of created being (Chapter 3). With the metaphysical structure of the world in mind we move to Maximus' thinking of the divine *energeia* (energy or activity) at work in the soteriological scheme of things (Chapter 4). The last chapter is devoted to the concept of participation which plays an important role in Maximus' total conception of the world in relation to God (Chapter 5).

The topic of participation is quite obscure, and I cannot think of any modern patristic scholar who has tried to define it in a more precise philosophical way. However, I believe I have found a key to the ancient understanding of the concept with the aid of a paper by Dominic O'Meara from 1980.

[1] *Amb.* 7, PG 91: 1084c–d: βούλεται γὰρ ἀεὶ καὶ ἐν πᾶσι ὁ τοῦ Θεοῦ Λόγος καὶ Θεὸς τῆς αὐτοῦ ἐνσωματώσεως ἐνεργεῖσθαι τὸ μυστήριον.

In the Conclusion I focus on two modern problems for which this Maximian cosmology is relevant: the human rights issue and the environmental issue. But although this book ends by highlighting the relevance of this ancient Christian system for the modern world and modern problems, what comes in between is a philosophical analysis and treatment of a fascinating, beautiful, but difficult philosophical and theological interpretation of created being.

I. ST MAXIMUS' WRITINGS AND THEIR INTERPRETATION

What can we say about the writings of this major thinker of high sophistication and penetrating mind? What kind of condition are they in, what kind of genres did he use and how readable are they?

In the *Alexiad*, Anna Comnena relates with admiration that her mother often was seen with a book by 'the philosopher and martyr Maximus'. One day Anna says to her mother that she herself would never have dared to listen to a doctrine as sublime as that of St Maximus'. His writings, Anna says, are *highly contemplative and theoretical* (*πάνυ θεωρετικόν τε καὶ νοερὸν*).[2] His *teaching*, obviously, is considered difficult to understand, and with this statement the modern interpreter can only agree. We could suspect also that the *texts* are difficult to read because of the complexity of their syntax, and to this the modern reader could bear witness as well. Philosophically hard, syntactically tough as they are, it is worth remembering that some of his writings are harder to read and understand than others, as we will see below.

It often helps to solve at least some initial riddles of a text if we are able to place it in a literary genre. When it comes to genres, however, one has to search carefully in the literature on Maximus to find substantial remarks.[3] This is unfortunate because the classification of a literary work according to its form often gives a clue to how it should be interpreted. Nevertheless, the consideration of genres will only make a difference if we know something of the author and

[2] Anne Comnène, *Alexiade* (Paris 1967), book 5, 9.3.

[3] Cf. Louth (1996), 20–2; Blowers (1991), ch. 1.

are able to place him within a certain historical setting. One could object, of course, that this is a rather awkward view of hermeneutics. On the other hand one should remember that the present subject matter is *Maximus' thought* and I find it reasonable to hold that the man himself and the context of his activity are both relevant for the interpretation of what he wrote.

What, then, do we know about him? I believe there are three relevant pieces of information that could help us: he had a solid education, became a monk and was engaged in the controversy over monotheletism. As a monk, his spiritual life developed within a certain setting, and as an *educated* monk we would expect his writings to reflect both a strategy to place monastic strivings within a wider soteriological scheme and a concern for exposing unsound spiritual doctrines for critique. Further, we should expect a coherent polemic against those he considered heretics. A first reading of the texts confirms initial anticipations and therefore makes it a reasonable project to search for the soteriological scheme and for the theoretical foundations of his polemics against the heretics. However, my purpose is not to investigate Maximus' anti-monotheletistic strategies, but to work out his philosophical ideas. On the other hand, a major thinker engaged in polemics of this kind, should be expected to work out a lot of categories of general philosophical relevance, and this he does. I am thinking of precise definitions of essence and nature, identity and difference, different kinds of unions, the concept of participation, etc.

In the study of ancient philosophy it is considered important to take into account Plato's dialogue-form and Aristotle's treatise-form. Andrew Louth classifies Maximus' works in three literary genres, which seems reasonable. First he mentiones the genre of gnostic centuries, further comes the genre of question and answer, and, finally, the commentary.[4]

Evagrius, Nilus, Diadochus of Photike, and Macarius practised the style of *gnostic sentences*.[5] Maximus' *Centuriae de charitate* (hereafter *De char.*) and *Capitula gnostica* (or *Capita theologica et oeconomica*, hereafter *Cap. gnost.*) are both in the genre of 'gnostic sentences'. Such sentences or chapters were combined in so-called 'centuries', each of

[4] Louth (1996), 20–1. [5] Thunberg (1985), 22.

which contains 100 small texts.[6] A 'chapter' is in fact not a very long section. It contains from one to just a few sentences. These chapters were arranged according to topic, as is seen for instance in the *De char.*, which contains four centuries *on charity*.

The genre was an established monastic style of writing before Maximus composed his works.[7] They invite the reader to slow and concentrated study, to some kind of 'contemplative reading'. Only this way would it be possible to detect the chains of meaning running through the text. In a monastic setting of Maximus' days the purpose of the genre would be to awaken practical understanding of the ascetic way with its internal connection to the deep truths of the Christian faith. Such texts are well suited for contemplative activity, hence the practical motive which lay behind the use of this genre. One would not gain much by reading through the text in one full swoop. Nor will a modern interpreter gain much if he is not going to carry the burden of contemplative reading. Rather, one has to dwell with the text, reflect over it and discover the connection between the chapters and centuries in a gradual uncovering of their central idea.

The writings composed as *questions and answers* were for the most part written to persons who posed different questions to Maximus. Here two things should be noted. Firstly, these works were commentaries on difficult texts from the Bible (*Quaestiones ad Thalassium*), from the Fathers (*Ambigua*) or both (*Quaestiones et dubia*). Maximus also composed plain commentaries, such as *Expositio in psalmum lix*, *Orationis Dominicae expositio* (hereafter *Or. Dom.*) and *Mystagogia* (hereafter *Myst.*). The last-mentioned work is an interpretation—rich in philosophical implications—of the Liturgy. Secondly, these writings are, according to the genre, connected with the monastic catechesis. They are mostly addressed to individuals and follow the pattern of question and answer that spiritual Fathers used when they instructed their disciples. Maximus has composed a work which directly anwers to this catechesis, the *Liber asceticus*, a dialogue between a spiritual Father and his pupil.

[6] In other authors there are examples of centuries with 90 and even 150 chapters. Louth (1996), 20, explains why for instance Evagrius uses 90 chapters.

[7] Thunberg (1985), 22; Louth (1996), 20.

The major works in the genre of question and answer, viz. the *Quaestiones ad Thalassium* (hereafter *Ad Thal.*) and the *Ambigua* (hereafter *Amb.*), are to a great extent composed as rather long and complex expositions of major theological and philosophical interest. The complexity concerns both the syntax and the theoretical contents. The language of the centuries is often easier and more readable than, for instance, the *Ambigua*. The *Ambigua* requires slow study and considerable linguistic and philosophical acuteness of its readers.

Several of Maximus' letters and of his *Opuscula theologica et polemica* (hereafter *Th. pol.*) also belong to the genre of question and answer.

The texts taken as the point of departure for my interpretation are partly found in Migne's *Patrologia graeca* (hereafter PG), volumes 90 and 91, partly in the *Corpus Christianorum, Series Graeca* (hereafter CCSG). In CCSG the most important critical editions are published. The series is, however, far from complete.[8] Moreover, new editions are being published at a slow rate. It could, of course, be considered a problem that the whole *corpus* has not yet appeared in a critical edition, but as it is we have no choice but to use the PG while we wait for something better. According to Lars Thunberg, a comparison between the PG edition and the CCSG edition of the *Ad. Thal.* shows that the PG is not inaccurate to a degree that should disturb us unduly. This is the verdict of Andrew Louth as well, when in a review essay from 1998 he says that (so far) the CCSG edition for the most part 'has provided welcome assurance of the general reliability of the text (of Combefis and Oehler) published in Migne's Patrologia Graeca (90–1)'.[9] We have no choice but to work with the text of the PG, since a full critical edition belongs to the future.[10]

II. ST MAXIMUS 'THE PHILOSOPHER'

Should we consider St Maximus a philosopher? This question in my opinion we should answer in the affirmative. If St Augustine is a philosopher, if Bonaventure is a philosopher, yes, if Plotinus

[8] See the bibliography for the editions I have used. [9] Louth (1998), 68.

[10] Thunberg has told me that he once (by letter) asked P. Sherwood about the condition of the texts, and received the answer that to a great degree we can trust the PG. I hope the future will not show that our trust was unfounded.

is a philosopher, Maximus is as well. He figures in the history of dogma as the greatest theologian of his century. He is remembered as the monk who defended the doctrine of the two wills of Christ against monotheletism. This defence, however, was based on a body of philosophical ideas developed prior to his engagement in that controversy. Like other Christian thinkers in the early Church he was not aware of any distinction in principle between Christian theology and philosophy. If we draw the line backwards in history, to St Justin Martyr, we can see how intellectual Christians could consider their own faith to be the true philosophy. This does not mean, of course, that Christianity, by its learned representatives, was looked upon as one philosophical school among other schools. But if Christianity is the *true* philosophy, then other philosophies would be disqualified as adequate interpretations of reality. Furthermore, as philosophy, Christian truth was deemed to have an intelligibility that legitimized it before the human intellect. In the *Dialogue with Trypho* Justin held that the task of philosophy is to inquire about the Divine.[11] Many have failed, he says, to discover the true nature of philosophy. Therefore a lot of different schools have emerged. Basically the science of philosophy is one and the same, and it was given to men for a specific reason. He says philosophy is precious in the sight of God, to whom it leads us and unites us.[12] A similar view of philosophy is found in St Gregory of Nyssa, when in *The Life of Moses* he defined *truth* as the sure apprehension of real being. The philosophical life is the life in quietness (ἡσυχία), which, on this view, was identical with the monastic life.[13] Philosophy, as the contemplation of divine matters, is not something that the Church has to borrow from 'the Greeks'. On the other hand, Gregory admits that Greek learning could be useful.[14]

Maximus was well aware that there existed a pagan or 'Greek' philosophy. But, on his view, genuine philosophy was Christian philosophy. Philosophy, in this understanding of the word, is taken as 'the love of wisdom', this *wisdom* being 'the Wisdom of God' (ἡ σοφία τοῦ Θεοῦ), which is the same as 'the Logos of God' (ὁ Λόγος τοῦ Θεοῦ), i.e. Christ.[15] In Christ, St Paul says, 'are hidden all the treasures

[11] *The Dialogue with Trypho*, ch. 1. [12] Ibid., ch. 2.
[13] *De vita Moisis* 2.23. [14] Ibid. 2.115.
[15] Cf. 1. Cor. 1: 24 and John 1: 1 ff.

of wisdom and knowledge'.[16] In the *De char.* Maximus says: 'The Christian philosophizes in these three things: in the commandments, in the dogmas and in the faith. The commandments separate the mind from the passions, the dogmas introduces it to the knowledge of beings, and faith introduces it to the contemplation of the Holy Trinity.'[17] Now, this saying is to be connected with Maximus' doctrine of a threefold spiritual development.[18] The terminology varies, but the first stage is in some passages called 'practical philosophy', the second is called 'natural philosophy', and the third 'theological philosophy'. We should not forget that the term philosophy has connotations in the direction of the monastic life, but this does not mean that in every instance we should simply equate philosophy with monasticicm: they may have the same reference, but not necessarily the same intention. To philosophize in the three things mentioned in the citation should be the normal activity of one living a monastic life or even of all Christians, as the text from *De char.* has it.

The Scriptures are not scientific or philosophical textbooks that explicitly teach the details of a philosophical world-view. On the other hand, the scriptural teaching about God and His creatures has implications that could be worked out philosophically. Further, on Maximus' view, God has revealed Himself, not only in the letters and syllables of the Scriptures, but in the created, natural world itself. There are two laws, the written law and the natural law, and both have equal honour and teach the same things. Neither is greater or less than the other.[19] The divine revelation in Scripture and nature is not something that is 'written out' in easily accessible articles. The basics are made explicit in the traditional dogmatic teaching of the Church, but even what this teaching implies for the details of a whole world-view has to be searched out by a properly trained intellect. An intellect is, according to Maximus, on its way to be properly trained when man moves on the path of spiritual development. A properly trained intellect is in this context the intellect of the good interpreter, and the good interpreter is the one who in a sound way is able to expound the different senses of the sources, i.e. the Scriptures and nature. Perikles Joannou says that for the Byzantines, philosophy is

[16] Col. 2: 3. [17] *De char.* 4.47, PG 90: 1057c.
[18] For details see the last part of ch. 4 § IV. [19] *Amb.* 10, PG 91: 1128d.

'Lebensanschauung'. It is neither a philosophy of concepts nor knowledge reached with the pure reason alone, but rather an experience of the All ('Ganzheitserlebnis') with the purpose of giving structure to human life and achieving likeness with God.[20]

Two qualifications should be made here. (i) Even though philosophy for Maximus is 'Lebensanschauung', he does not lack interest in theoretical *concepts*. Rather, he tries to work out definitions of several important, ontological conceptions, for instance the meaning of different kinds of unions, the meaning of essence, nature, hypostasis, *enypostaton*, will, etc.[21] Some of his lists of definitions or treatments of terms could remind one of Aristotle's *Metaphysics* book 5. (ii) Even though Maximus is not preoccupied only with the kind of knowledge we gain with the pure *ratio*, he seems to think that the Christian faith somehow is intelligible. Here we move, however, into a rather intricate subject area, for according to Maximus the essence of God, the triune character of the divine being, its internal life and divine activities, are (according to their inmost nature) beyond human comprehension. Whenever we talk of God, philosophically or popularly, we *may* talk truly if we talk in accordance with right reason, but our talk can never be adequate. We are constrained to talk within the boundaries of created being, but what we talk of (i.e. God) is not limited by the categories of created being. There are mysteries transcending the human mind, but transcending the capacity of the mind does not mean that something is in itself contrary to reason, i.e. unreasonable, but that it is beyond human capacity to reach it.

Let us return to the quotation from *De char*. 4.47. The Christian philosophizes in the dogmas, and the dogmas introduce the mind to knowledge of beings, Maximus says. Even though not immediately intelligible, I think this may be a convenient starting point to develop some thoughts on an important philosophical subject, viz. that of *method*. 'The dogmas introduce the mind to knowledge of beings.' Now, a twofold question emerges here: what is meant by dogmas and how do they introduce knowledge of beings? A quotation from *Mystagogia* chapter 18 may offer a glimpse into how Maximus thinks

[20] Joannou (1956), 2.

[21] Cf. *Th. pol.* 18 (PG 91: 213a–216a), 23 (PG 91: 260d–268a), and 26 (PG 91: 276a–280b).

in this connection: 'The profession by all of the divine symbol of faith signifies in advance the mystical thanksgiving to perdure through the age to come for the marvellous *logoi* and *tropoi* by which we were saved by God's all-wise Providence on our behalf.'[22] We should note the liturgical context: Maximus comments on the Divine Liturgy. The profession of the Creed signifies the thanksgiving for the *logoi* and *tropoi* 'by which we were saved'. On Maximus' view, that is, the Trinitarian confession indicates the cosmological and soteriological system that he expounds in his philosophical theology. His doctrine of *logoi* and *tropoi* are in fact, as we shall see, an interpretation of the Christian faith in which the system of reality is presented as a divine device to accomplish the deification of man and nature.

Basic principles of his system, then, are found in the liturgical insight into dogmatic truths, and from these principles he deduces philosophical implications. The basics of his system are the credal confession to the divine Triad and to God's Incarnation as it is understood by the Council of Chalcedon (451): the uncreated and the created nature in Christ are united hypostatically without confusion, without change, without division, without separation. This 'logic', according to Maximus, not only regulates the primary instance of the Incarnation of the Logos, it is also a general law for the relation between the divine and the created sphere.[23] From these principles a rational interpretation of being (the world and man in relation to God) is built up by the aid of traditional logical procedures. The interpretation is expressed in a critically revised and Christianized philosophical vocabulary.

His philosophical terminology, even though largely identical with that of the philosophical schools, is filled with a new, Christian content. Maximus has a philosophical vocabulary that he did not *borrow* from outside of the Church, even though the Neoplatonists used a lot of the same terms. From the time of the great Cappadocians (fourth century), through the centuries of Christological controversy,

[22] *Myst.* PG 91: 696a–b.

[23] Törönen (2007), in the introduction to his book, objects to this 'pan-Chalcedonianism' and points to union and distinction as basic logical conceptions of Maximus' thinking. I do not object to this, but I still think that the 'Chalcedonian logic' could be seen as an expression of a first concern with union and distinction. The end result will not differ in any important respect.

Christian authors had taken over and 'Christianized' many philosophical terms which by the time of Maximus were established as a normal Christian vocabulary. He received a Christian intellectual heritage that could freely express itself in this kind of vocabulary, and strictly speaking, these are not 'Neoplatonic terms', rather they are Greek words, used by the Fathers. And if we would like to know what these words mean to them, we should search their own writings and not those of the philosophers, their own context of discussion and controversy and not those of others. This does not mean, on the other hand, that it is futile to study the philosophical schools when working on the Fathers. Whenever the philosophers and the Fathers put forward differing and even rival theories it is useful to search out why they do so. What are the motives behind the differences? Why do Christian thinkers develop alternatives to pagan doctrines? Asking such questions one may discover the reasons behind differences and likenesses in ancient thought. If there is any 'influence' going on at all it could only mean that someone read or heard what another wrote or taught, and accepted it as a good idea, not that he succumbed to some kind of intellectual virus.

When it comes to method as such, the first question to be addressed is *how*, according to Maximus, one gains knowledge of first principles. This question seems to have a twofold answer. First we have to say that knowledge of first principles is initially gained through what we could call the *ecclesial or liturgical experience*. This experience is the common experience available to all in the Christian Empire and formative of their world-view. Here we could once more point to the citation from *Mystagogia* 18: 'The profession by *all* of the divine symbol of faith ...' This happens within the Liturgy as an expression of the traditional faith of the assemblage (ἡ ἐκκλησία). What is professed is what is lived or experienced by the faithful. It is not a piece of knowledge that is special or under pressure from sceptical minds, rather it is the normal or common world-view. Secondly, genuine insight into or spiritual knowledge of first principles is gained through spiritual development.[24] As I have said, Maximus speaks of a threefold way through practical philosophy, natural philosophy, and mystical theology. The point is that to be able to see the cosmos

[24] For details, see Ch. 4 § iv

in light of its basic principles, man must develop the virtues, gain detachment, and become free from worldly cares and temptations. In this process the mind becomes simplified, according to Maxmus, and in the higher forms of natural philosophy or contemplation the whole cosmos is seen as an ordered whole based on the divine principles instituted by the Holy Trinity.

As in the case of several other philosophers, Maximus thought often moves intuitively from premises to conclusions. All the premises that substantiate a conclusion are not always brought forward, and the inference is drawn in an intuitive leap without the argument being formalized as a syllogism or a propositional logical scheme. It may happen, though, that the formalized scheme be discovered just below the surface of the argumentation, as can be seen, for example, in his arguments for the temporal beginning of the world. In the tenth *Ambiguum*, Maximus presents an argument from motion.[25] It is easily reconstructed even though it is not entirely worked out in the text: everything that is in motion has a beginning, because everything that is in motion has a cause, and everything that has a cause has a beginning. As will be seen in Chapter 2, this, and its further implications, may be formalized in a syllogism.

In *The Cambridge History of Later Greek and Early Medieval Philosophy* Maximus is treated under the heading 'The Greek Christian Platonist Tradition from the Cappadocians to Maximus and Eriugena'.[26] It is tempting to label Maximus' philosophy 'Christian Neoplatonism'. Is this label justified? It is a common view that Thomas Aquinas consciously worked for a synthesis between traditional Western Christian thought and Aristotelian philosophy. One could not say however, that Maximus consciously made an effort to accomplish a synthesis between Christianity and Neoplatonism. As a label used in histories of philosophy the term 'Christian Neoplatonism' sounds to my ears a bit suspect. One should always keep in mind that even if Neoplatonist doctrine and the Christian thinking of the Greek Fathers show similarities and may be fruitfully compared, the Christian philosophy of these Fathers is an autonomous body of thought that in many instances differs basically from Neoplatonism. To repeat what

[25] *Amb.* 10, PG 91: 1176d–1177b.

[26] Armstrong, ed. (1980), 421 ff.

was suggested above: Christian theologians are not passive victims of influence, as if they do not think; rather they themselves philosophize.

I believe that Maximus is a genuine philosopher, not so much because he used a certain vocabulary and employed rational conceptualizations, but because he was able, on the basis of liturgical experience and dogmatic insight, to use this heritage in a creative way, to think through and express systematically and (at least seemingly) coherently the implications that Liturgy and dogmatics have for a Christian cosmology, metaphysics, and ontology.

Maximus has made an original philosophical contribution because nobody else before him—or perhaps even after him—has systematized the implications of Christian teaching and practice in the working out of a world-view that to the same degree is centred in Christ, the Logos of God. His writings are not written as systematically arranged philosophical treatises. There are sections in his writings which are of a more or less direct philosophical nature, but more importantly, in my opinion, is the fact that virtually all he has written reflects a philosophical interpretation of the world in its relation to God. His Christian philosophical outlook works in the background of his whole doctrine then, and it is possible to reconstruct this philosophy from what he explicitly says in his texts or from implications from the texts. Such a reconstruction is what I am trying to accomplish.

III. ST MAXIMUS' PHILOSOPHICAL SOURCES

I suppose it looks a bit strange that, after what was said about philosophical influence in the previous paragraph, there follows a paragraph on philosophical sources. However, there is no reason for disturbance. It seems quite obvious that St Maximus knows philosophical works from the pagan schools and acknowledges some of their doctrines. What I deny is that he can be reduced to being simply the sum of alleged 'sources' as if he were simply a recipient or that such sources can tell us anything about his doctrine. For instance, if it could be demonstrated that he knew Neoplatonic or Aristotelian material it would still be wrong to burden Maximus' understanding of essence (*οὐσία*) with Neoplatonic or Aristotelian tenets. What he himself teaches must be searched out in his own texts, within his own tradition and within the polemical situation he is in.

That Maximus is well versed in the theological traditions is easy to show. He quotes many authors by name and discusses texts that are drawn from several of the great Eastern Fathers. He formulates his thought with due consideration of the sources of the mainstream orthodox thinkers and with a polemical attitude towards heretical views. Certainly, much of his philosophical inspiration is from the theologians, for instance from Basil the Great, Gregory of Nyssa, Gregory the Theologian, Nemesius of Emesa, Cyril of Alexandria, Dionysius the Areopagite, Leontius of Byzantium, and Leontius of Jerusalem; what is more difficult to determine is which philosophical works he read. As I said above, he does not have to move outside of his theological tradition to find the greater part of his philosophical vocabulary; but, on the other hand, it seems obvious that his knowledge of *other* philosophies goes beyond what he can extract from the Fathers.

According to the *Vita et certamen* he received the full academic training of his time.[27] The problem here is that this *Vita* is composed by a Studite monk in the tenth century, and even if it relies on earlier material, we are left in uncertainty about Maximus' early years.[28] Now, if Maximus received this form of education he would, according to Sherwood, have begun his training in his sixth or seventh year and continued to his twenty-first.[29] Sherwood says about this training that it comprised elementary disciplines like grammar (i.e. the reading and study of the classical authors), rhetoric, and philosophy. Philosophy included arithmetic, music, geometry, and astronomy, and it is said to comprise logic, ethics, dogmatics, and metaphysics. 'The instruction was based on the writings of Plato and Aristotle with the commentaries of Proclus, Iamblichus, Alexander Aphrodisiensis, Ammonius, and Porphyry.'[30] After stating this, Sherwood continues: 'One cannot say for certain that Stoic and Neoplatonic doctrine served directly as material for instruction.' Maximus could have gone through a curriculum of this sort, but if we should try to determine the range of his knowledge of specific authors and texts from his

[27] *Vita et certamen*, PG 90: 69c ff.
[28] Cf. Louth (1996), 4.
[29] Cf. Sherwood (1950), 387–8, and (1952), 1–2.
[30] Sherwood (1950), 348.

writings, we could not substantiate Sherwood's claims. I am not able from his writings to say whether he had any direct acquaintance with this or that philosophical text by this or that author. For instance, the concept of *ousia* (essence) is central to his thought, but we cannot say that it builds on Aristotelian sources or that he knew specific Aristotelian texts—naturally, for a term so central to Christian debates, Maximus did not have to resort to any theoretical discussions outside the tradition. And, while Maximus is acquainted with the kind of logic that stems from Aristotle's *Categories*, it is not even possible to say for sure that he knew Porphyry's *Isagoge*. He *could* have read Porphyry, but the kind of logic extracted by the Neoplatonists from the *Categories* was, as we shall see (especially in Chapter 3), well known by the Fathers, and Maximus could acquaint himself with it from them. However, a hypothesis could be constructed that Maximus in addition knew this kind of logic from logical compendia and textbooks from the seventh century. Mossman Roueché has, in two important articles, described four and published three such texts.[31] Two of them occur in manuscripts attributed to Maximus himself, and about these Roueché says:[32] 'as those attributed to Maximus are preserved exclusively in mss containing his genuine works, the likeliest explanation for the attribution is that they were found among his papers after his death (662 A.D.) and mistakenly transmitted in his name.' An alternative hypothesis to that of Roueché could be that later editors of Maximus' genuine works considered these logical texts a useful tool in understanding him. If Roueché is correct, however, we are lucky to have *some* philosophical material to investigate which Maximus himself actually knew. One of these texts is a logical compendium and another one is a handbook. They contain and explain some of the technical vocabulary of Aristotelian–Neoplatonic (Porphyrian) logic. One of the texts contains a treatment of the Aristotelian categories and an exposition of the Porphyrian tree as well.

In addition to these texts I would like to frame a hypothesis about another philosophical aquaintance that Maximus could have made. The Christian Neoplatonist Stephanus, a pupil of John Philoponus in Alexandria, is held to have moved to Constantinople when Heraclius

[31] Roueché (1974) and (1980). [32] Roueché (1974), 63.

became emperor in 610.[33] Stephanus was the author of commentaries on Aristotle, for instance on *De anima III* and *De interpretatione*. In Constantinople he is held to have lectured on Plato and Aristotle, on geometry, arithmetic, and astronomy.[34] Now, *if* Stephanus actually moved to Constantinpole in 610, and *if*—as he probably did—he brought with him his own in addition to other books on philosophy (by the Christian Neoplatonists Elias and David?), and *if* he gave lectures, and *if* Maximus was made head of the imperial Chancellery by Heraclius about 610,[35] St Maximus would be in a position to join the learned man from Alexandria, listen to his lectures and read his books. It seems to me, however, that there is not much to be gained from our knowledge of such a connection. In my opinion, there is no major need to resort to these writings—however interesting—in order to establish a proper context for reading Maximus, or for finding clues to interpret his thought.

IV. EARLIER RESEARCH RELEVANT TO THE PRESENT TOPIC

Almost all those who have written anything about St Maximus make comments on his cosmology or metaphysics but, since only a few of them have looked at his thought from a philosophical point of view, the philosophical structure of his system and its basic principles have not been sufficiently investigated. Eric Perl, however, is an honourable exception. In his 1991 dissertation *Methexis: Creation, Incarnation, Deification in Saint Maximus the Confessor*, he moved into at least two important topics from an explicitly philosophical point of view: the Porphyrian tree and the participation-problem. Perl emphasizes

[33] Armstrong, ed. (1980), 483. The hypothesis that Heraclius summoned Stephanus to Constantinople is rejected by Beck in 1966, but supported by Lumpe (1973), cf. Kazhdan ed. (1991), 1953. Further it is accepted by Blumenthal in 1976 and by Sorabji in 1990, cf. Sorabji, ed. (1990), 311 and 16. But according to Wilson (1983), 47, 'almost every sentence that has been made about him is open to doubt'.

[34] Armstrong, ed. (1980), 483.

[35] Törönen (2007), 14 n. 1, thinks it probable that Maximus was head of the imperial Chancellery, but not that his title was *protoasecretis* (cf. Louth (1996), 4–5) since this title (Törönen remarks) emerged in the middle of the 8th century.

the importance of the *logoi*-doctrine for Maximus, and shows how the different aspects of his system form an integrated whole. While I disagree with Perl on some specific points as will be seen below, nevertheless, his thesis represents a major contribution to the Maximus literature.

It might be convenient to distinguish between the older, mainly post-war contributions, and the more recent research on Maximus at the turn of the twenty-first century. Significant contributions are made by von Balthasar's *Kosmische Liturgie* (1941 and 1961),[36] with its vision of the totality of St Maximus' thought; Polycarp Sherwood's *The Earlier Ambigua of Saint Maximus the Confessor and his Refutation of Origenism* (1955), with its stress on the anti-Orgenism of the *Ambigua*; and Lars Thunberg's *Microcosm and Mediator* (1965), which is a penetrating study of St Maximus' anthroplogy.

Belonging to the phase of Maximus scholarship just before the end of the twentieth century, is the highly valuable study of the *Ad Thal.* by Paul Blowers, *Exegesis and Spiritual Pedagogy in Maximus the Confessor* (1991). (Blowers and Robert Louis Wilken have also rendered the English-reading public great service with a translation of important texts from Maximus' works, published in 2003 as *On the Cosmic Mystery of Jesus Christ.*) A study of importance to any future investigation of the topic of deification is Jean-Claude Larchet's *La Divinisation de l'homme selon Saint Maxime le Confesseur* (1996). However, there is one major problem in Larchet's interpretation of Maximus: his denial that the theory of participation plays a systematic role in the Confessor's thought. According to Larchet, Maximus does not develop a precise doctrine of participation even though he occasionally uses the terminology.[37] As will be seen from Chapter 5 below, I believe that Larchet is wrong. I will argue that the whole metaphysics of Maximus, including his doctrine of creation and deification, would lack its keystone without such a concept.

In 1996 Andrew Louth's *Maximus the Confessor* appeared with translations of some important texts and a good introduction. For the first time whole parts of the *Ambigua* appeared in English translation

[36] A translation into English by Brian E. Daley was published in 2003, see bibliography.

[37] See Larchet (1996), 600–1. Cf. 601 n. 305.

and this did much, I am sure, to stimulate the interest in Maximus' thought.

Despite the contribution to Maximus scholarship during the last half-century, this literature suffers, nevertheless, from a neglect of the Neoplatonic background of the Confessor's thought—though Perl's study is here a notable exception. I confess this with some reluctance, for as I said above I am sceptical about the strategy of interpreting Christian thinkers in the light of non-Christian philosophy, as if they could be reduced to 'Platonists', 'Aristotelians', etc. To quote from Mark Julian Edwards' stimulating book on Origen, scholars often seems to hold that 'a Christian never thinks but only inhales the thoughts of others'.[38] For instance, when von Balthasar writes about Maximus' basic concept of *ousia* he starts his treatment with the Aristotelian distinction between primary and secondary substance.[39] Thunberg, following von Balthasar, says that the duality of Maximus' concept of essence probably goes back to Aristotle.[40] As will be seen in Chapter 3, I believe that it is misleading to put so much stress on the Aristotelian roots of this important concept. We should have to ask: did Maximus really need any pagan antecedent to acquaint him with a sophisticated doctrine of essence?[41] There is an obvious answer to this question: not at all. Maximus only had to study the Christian contributions to the Christological controversies to find the strategic tracks he would have to move in accordance with. 'He would have to move in accordance with'—I realize these words sound rather passive, as if a new source for the virus of influence should be substituted for the old one. This is not, however, what I have in mind. On the one hand, as an educated monk living in a mainstream Christian tradition, he possessed a hermeneutical key to the writings of others; on the other hand, we should remember that a lot of this literature originated from people who believed they could solve urgent problems. Maximus, for his part, thought he could see how orthodox dogma fitted into a general philosophical framework, nurtured from what he considered the tradition of 'Fathers'. We should resist the temptation to reduce his achievement to the influence of historical agency—if he is *conscious* of being and thinking within a tradition. It is precisely the

[38] Edwards (2002), 54.
[39] Von Balthasar (1961), 213 ff.
[40] Thunberg (1995), 83.
[41] Cf. Edwards (2002), 8.

consciousness of such a person that makes him an original thinker, not the kind of person who just repeats his predecessors. When I speak of 'the Neoplatonic background' in this context, it should not be stressed further than this: in Neoplatonic circles one was thinking in patterns that came close to the doctrines and philosophy of the Christians. And therefore something might be learnt from the pagans, if one should happen to be acquainted with the relevant texts. In order to follow up on the doctrine of *ousia*, it *could* be fruitful to note that Maximian thought reminds one of the Platonic idea of *ousia* as the highest inclusive 'category' with its roots in the 'greatest kinds' in Plato's *Sophist*.

Perl, seeing the importance of the philosophical background, points to Stephan Gersh's book *From Iamblichus to Eriugena* (1978) as providing valuable material for the study of St Maximus.[42] To some degree I agree with Perl on this. However, I feel that the gain of studying Gersh is limited because of the special approach he takes to his topic. Perl's verdict is illuminating here:[43] 'Gersh's approach, however, is analytic rather than synthetic, as he isolates and examines certain themes or even terms in his subject's works but does not attempt to present any of their theories as a unified, coherent ontological structure.' Thus, the value of Gersh's study is somewhat limited.

Before I turn to more recent work, there are some other publications of importance for the study of the philosophical context of St Maximus. These include Richard Sorabji's *Time, Creation and the Continuum* (1983) and *Matter, Space and Motion* (1988). Sorabji develops certain philosophical themes from Antiquity and places great stress on the Neoplatonist commentators. He even moves into the field of the Church Fathers and shows how they cope with certain physical and metaphysical topics in tension with the pagan traditions. Another source of material, associated with the name of Sorabji, is the series of translations from the Neoplatonic commentators. The first volume was published in 1987.[44] *The Ancient Commentators on Aristotle*, with Sorabji as general editor, was originally planned to

[42] Perl (1991), 11. [43] Ibid.

[44] Philoponus, *Against Aristotle on the Eternity of the World*, trans. Christian Wildberg.

include forty volumes. All this material has been available, of course, in the original Greek in the CAG, but I am sure that the publication of translations will give scholars the oportunity to aquaint themselves more easily with the kind of material offered by these important sources. The work of the commentators, whether Peripatetics or Neoplatonists, is an important element in the intellectual climate of the centuries before Maximus, and the legacy of the pagan schools is still present in his time.

In 2000 I defended my thesis *The Christocentric Cosmology of St Maximus the Confessor* for the degree of *Doctor Philosophiae* at the University of Oslo. Copies of the text were distributed to scholars both in Europe and in North America. Since then several new and important works of Maximus have appeared, some of which mention or take notice of my research: Assad E. Kattan's thesis *Verleiblichung und Synergie* (Leiden: Brill, 2003); Demetrios Bathrellos, *The Byzantine Christ: Person, Nature and Will in the Christology of St Maximus the Confessor* (Oxford: Oxford University Press, 2004); Pascal Mueller-Jordan, *Typologie spatio-temporelle de l'Ecclesia byzantine, La Mystagogie de Maxime le Confesseur dans la culture philosophique de l'Antiquitι tardive* (Leiden: Brill, 2005); Adam G. Cooper, *The Body in St Maximus the Confessor* (Oxford: Oxford University Press, 2005); and Melchisedec Törönen, *Union and Distinction in the Thought of St Maximus the Confessor* (Oxford: Oxford University Press, 2007). All these books highlight different aspects of the Confessor's thought, some of which have proved stimulating to my work on the present revised version of my thesis. This, in a special sense, is true of Törönen's book, which challenges some of the views I put forward in my thesis.

All these publications witness both to a need and to a will to move beyond major interpreters like von Bathasar, Sherwood, and Thunberg. It is promising too that more young scholars are about to start up new projects in the field.

2

The Divine Ideas and the Creation of the Cosmos

I. 'EXEMPLARISM'

St Maximus' doctrine of divine *logoi*, or—to be more precise—of *λόγοι τοῦ εἶναι*, is a kind of doctrine of Ideas. This is how I understand the important text of *Ambiguum* 7, where he says that God, before the ages, possesses *logoi* of everything that has been made through the creative act. God brought forth from non-being both the visible and the invisible creation by these *logoi*. The expression *ἡ ὁρατὴ καὶ ἀόρατος κτίσις* denotes the two regions of the created cosmos, the sensible and the intelligible. The expression is meant to be exhaustive for the cosmic totality, and embraces everything that in any way is created by God. The words are probably based on the symbol of faith.[1]

There is, according to Maximus, a *logos* of angels and of 'every essence and power filling the world above'. There is a *logos* of man, and of everything that receives its being (*τὸ εἶναι*) from God.[2]

All beings, consequently, have been created from divine *logoi* or, in a more Platonic way of speaking, from divine Ideas. Historians of medieval philosophy have called the doctrine that the world is created from divine Ideas 'exemplarism'. This term is used, for instance, in F. Copleston's *A History of Philosophy* to denote the philosophical doctrine of God as *causa exemplaris*. As *causa exemplaris* God

[1] *πιστεύομεν εἰς ἕνα Θεὸν πατέρα παντοκράτορα, ποιητὴν οὐρανοῦ καὶ γῆς ὁρατῶν τε πάντων καὶ ἀοράτων.*

[2] *Amb.* 7, PG 91: 1080a.

possesses in His intellect all the Ideas which together make up the pattern of the created world.[3]

I think we should use the term exemplarism also to include the traditional Platonic doctrine of Ideas as patterns. Exemplarism in Plato is primarily connected with the *Timaeus* and with the interpretation of this dialogue in the diverse Middle Platonic and Neoplatonic systems. It was discussed by early Christian thinkers—among others by St Augustine—and received much elaboration in Western Christian thought.

It is tempting to say that nothing could be more 'Platonic' than exemplarist doctrines. But what, I would ask, is the *Platonic* character of such doctrines? Well, there is a doctrine of Ideas and of things participating in or resembling these ideas, would not that be a Platonic doctrine? More needs to be said about this.

As a general rule I think exemplarist doctrines in Christian 'systems' should be termed 'Christian'. If one uses the term Platonic, it should just be in the limited sense of 'a doctrine resembling a Platonist doctrine'. I think exemplarism was developed within a Christian context in order to answer a cosmological challenge confronting Christian thought. Maybe the authors of the theory knew that similar theories existed within other systems as well, and maybe they even learnt something from non-Christian thought. That does not mean, however, that a whole Platonic metaphysics in a mystical manner made its way into Christianity and effected a change in its whole mentality. This is not a case of a 'Platonization' of Christianity.

Christian exemplarism was developed as an element integral to the Christian understanding of reality. From the opening words of Holy Scripture one learned that God, in the beginning, created heaven and earth, sea and dry land, all kinds of plants, heavenly bodies, life in the seas, on dry land and in the sky, and, finally, man. Creation is described as an ordered system, a *kosmos*. Within this cosmos, God created all living beings, each one according to its kind (κατὰ γένος),[4] each one in its natural place in the cosmic building. For Christian thinkers, implicit in the description of creation in the book

[3] Cf. the index in *A History of Philosophy*, vol. 2, part II, 'exemplarism', with several references to Augustine, Bonaventure, Thomas Aquinas, and Duns Scotus.

[4] Cf. Gen. 1: 11–12 and 21 (LXX).

of Genesis, was the presupposition that the world resulted from a divine plan or design. In Proverbs (ch. 8) one could read that Wisdom arranged or ordered (ἁρμόζουσα) the world when it was established. The Wisdom of Solomon (11: 20 LXX) tells that God ordered everything according to measure, number, and weight. In Ecclesiasticus (16: 26 LXX) it is said that the works of the Lord were in His council from the beginning (ἀπ' ἀρχῆς), and in the creative act He divided their parts, which could mean that He ordered the world according to plan. For a Christian interpreter the term *archē* naturally leads the thoughts to Him who was 'in the beginning' (ἐν ἀρχή), the Logos of God, Christ (cf. John 1: 1). Consequently, the apostle Paul could write (Col. 1: 16–17) that in Christ 'were all things created,... all things were created by Him, and for Him; and He is before all things, and by Him all things consist'.

It was quite natural for Christian thinkers in a Hellenistic context to seek to formulate such basic Christian insights in a suitable philosophical terminology, thereby giving expression to a Christian view of the world. To the Christian thinkers, this was not just a question of apologetics (to legitimize Christianity as a rational phenomenon), but was bound up with understanding the scriptural message. Wanting to explain how everything comes to be *from Him* (ἐξ αὐτοῦ), is established *through Him* (δι' αὐτοῦ), and has its goal *in Him* (εἰς αὐτόν) (Rom. 11: 36), it seemed reasonable to do this by showing how the total cosmic order is kept within the divine Providence in such a way that God has in His possession the plans of everything.

If this genuine Christian idea is given precision in suitable philosophical terms, it will not be any less Christian because of it. The basic motif of Christian exemplarism is, consequently, neither Platonic, nor Neoplatonic, but originates from the Christian understanding of God and His creation.

II. PRECURSORS TO EXEMPLARIST DOCTRINES

The Cosmology from *Timaeus*

The immense importance of the dialogue *Timaeus* for cosmological thought up to the Renaissance is well known. In the *Timaeus* we find, among other things, a Platonic cosmology.[5] Plato says that everything that has come to be necessarily must come from some cause (ὑπ' αἰτίου). This holds both for the makings of the artisan, and for the cosmic order. The cause of the cosmic order is presented under various names. It is called god, creator, father, and—most famously—demiurge.[6]

The Demiurge is spoken of in such a way that one gets the impression that this is some kind of personal, divine power, much different from the 'self-thinking thought' of Aristotle's *Metaphysics* book Λ. The motivation behind the creativity of the Demiurge is his *goodness*, his wish to create as good a cosmos as possible. By setting his attention on the *paradigm* he establishes order in a previously unordered matter. This paradigm is eternal (ἀΐδιον), and is described as a 'living being' (ζῶον), a living being which embraces all the intelligible living beings (τὰ ... νοητὰ ζῶα πάντα). There seems to be common agreement among interpreters that the paradigm or living being is the ideal pattern of the world, the unity of all Ideas for living substances which fill the cosmos when it is put in order.[7] In *Timaeus* 39e–40a there is a list of four main genera which belong to the paradigm, viz. 'the heavenly kind of gods; another the winged kind which traverses the air; thirdly, the class which inhabits the waters; and fourthly, that which goes on foot on dry land'. Cornford comments: 'These main types, as well as the indivisible species of living creatures and their specific differences, are all, in Platonic terms, "parts" into which the generic Form of Living Creature can be divided by the dialectical procedure of Division (διαίρεσις).' In my opinion, however, Cornford is here overreaching his conclusions. We know of Plato's skill in the diairetic art from many dialogues (for instance the *Sophist*), but that he should be thinking here of a veritable Porphyrian tree of genera and species seems a bit anachronistic.[8] I believe it is in later Platonism, inspired by the Aristotelian classifications of living beings, that the Porphyrian tree is connected with the paradigm. It is probable that Plato in the *Timaeus* has had certain ideal types in

[5] My interpretation is mainly built on *Tim.* 27d–30d.

[6] See Guthrie (1986), 253 n. 2 for references. The term 'demiurge' is the most well known, but the term 'god' is the one most frequently used.

[7] Cf. Guthrie (1986), 255–9. See *Tim.* 30c–d.

[8] On the Porphyrian tree, see more below.

mind, but hardly a complete taxonomy of everything living. We shall return to this subject later, since in Neoplatonism the divine Ideas are ordered according to a taxonomic system, and taxonomy plays a role in St Maximus' thought as well.

The cosmology of the *Timaeus* represents the most developed exemplarism that we can find in Plato's writings: the Demiurge creates the cosmos while contemplating the paradigm. But the world-plan that he contemplates is not Ideas *within* his own thought. In relation to the Demiurge, the Ideas are 'outside' him. They constitute an external object for his consideration.[9] The cosmology of the *Timaeus* is an important step along the way that leads on to Middle Platonic and Neoplatonic exemplarism.

Aristotle's Doctrine of Categories, Definitions and Classifications

Aristotelian philosophy plays a role in the transformation of the theory of Ideas which finally culminates in the exemplaristic doctrines of Late Antiquity. Everyone knows that Aristotle criticized the Ideas.[10] But the Socratic–Platonic notion that there may be established *ideal types* representing sensible phenomena is important for him.[11] It should, therefore, be possible to establish such types for the species and genera of living beings. These may—with the aid of Platonic division (διαίρεσις)—be organized in a systematic, hierarchic 'map' of reality, especially of the animal kingdom. But according to Aristotle, universal species and genera cannot exist by themselves, but only as abstractions from the horizontal order of particular existents.[12]

In the *Categories* Aristotle considers how specific and generic concepts may be predicated, for instance:[13]

[9] Not everyone has agreed on this interpretation. Some interpreters have seen the paradigm as the thoughts of the Demiurge. The text of the *Timaeus* should, however, be clear enough, so that such an interpretation seems forced. Cf. Guthrie (1986), 259 ff.

[10] *Metaph.* A, ch. 9.

[11] Plato, *Republic* 596a; Aristotle *Metaph.* M, 4. 1078^b32 ff.

[12] *Metaph.* M, ch. 4: 1078^b30 ff., *Cat.* 5. 2^b3 ff. and *An. post.* B, ch. 19 about the abstraction of universal concepts.

[13] Cf. *Cat.* 1^b10 ff.; 2^a11 ff.

(i) Socrates is a man.
(ii) A man is an animal.
(iii) An animal is a substance.

Between (i) and (iii) a great many predications may be conceivable, viz. exactly as many as will correspond to the natural order of living beings. But, according to Aristotle, in every case there will be a *finite* number of predications; that is, a finite number of specific and generic terms, between the individuals (primary substances) and the highest category itself (οὐσία).[14] Many texts in the Aristotelian corpus show that what has later come to be labelled a 'Porphyrian tree', is already present in its primitive form in Aristotle himself. He has seen the possibility that living creatures may be classified in a hierarchic system of species and genera with the particulars as starting points.[15] As we have already seen, such a system does not constitute a really existent realm of Ideas in the vertical order, but is an abstract system of concepts which has its ontological basis in the horizontal realm of individualized substances. As we shall see, however, this status changes in Neoplatonic interpretations. What is important for the history of exemplarism, is the Aristotelian notion of a possible systematic and conceptual mapping of beings. This Aristotelian transformation of the doctrine of Ideas was destined to play a role in the exemplaristic thought of Late Antiquity.

The Stoic Doctrine of Logos/*logoi*

The last theory to be mentioned is the Stoic doctrine of *rationes seminales* (λόγοι σπερματικοί). To the Stoics divinity was held to be an immanent, material force, a creative fire or first fire (πῦρ τεχνικόν or τὸ πρῶτον πῦρ).[16] As such, the divinity contains all the *rationes seminales*, which make everything happen according to fate.[17] The Stoic divine *logoi* are unitary forces, which, by analogy with animal

[14] Cf. *An. post.* A, 22. 83^{b}1 ff.

[15] E.g. the discussion in *De partibus animalium* book 1, chs. 2–4.

[16] Cf. Long and Sedley (1992), 1: 274/2: 271, A Aetius; 1: 275/2: 273, D Stobaeus; 1: 276/2: 274, G Aristocles.

[17] Long and Sedley (1992), 1: 274/2: 271, A Aetius; 1: 275/2: 272, B Diogenes Laertius.

seed, potentially contain the growth and development of the things that originate from them. They serve as principles for everything, and are the main causes of what happens by nature.[18] Everything that happens in the cosmos unfolds successively from the potential of the *logoi*, and the beings that come to be during the whole course of world-history are successively instituted from *rationes seminales* as principles.

These three elements, respectively from Plato, Aristotle, and the Stoics, contributed to the development of exemplaristic thought in Middle Platonic and Neoplatonic philosophy.

III. EXEMPLARISM IN THE PLATONIC TRADITION

A lot of differing systems of thought, some known only from the report of others, are grouped together under the label Middle Platonism. Antiochus of Ascalon (*c.*130–*c.*68 BC) is held to be one of the founders of this school—if it is appropriate to talk of a school and of a founder.[19] According to Dillon, Antiochus' God reminds one of the Demiurge of Plato and the Stoic immanent divinity.[20] The paradigm of the created cosmos is, Dillon suggests, contained within the Demiurge or Logos. This paradigm, which is the pattern of the created cosmos, is the sum total of God's *rationes seminales*. In Dillon's opinion, we have here for the first time the doctrine of the Ideas as God's thoughts.[21]

In Plutarch (*c.*AD 45–*c.*125) we find the Ideas contained in a transcendent principle called the Logos, which is the medium through which God relates to the material cosmos.[22] The theory of Ideas as thoughts or designs of a divine mind is a *cosmological* doctrine.

[18] Long and Sedley (1992), 1: 276/2: 274, G Aristocles.

[19] According to Dillon (1977), 84, some elements of Antiochus' thought indicate that he is not to be regarded as the sole father of the movement. In my treatment of Middle Platonism I am for a great deal dependent upon Dillon's study, which is held to be the standard work on the subject.

[20] Cicero, *Acad. post.* 28–9 and Dillon (1977), 82–3.

[21] Dillon (1977), 95. It is possible that Antiochus was the first to teach this, but we cannot be sure.

[22] Ibid. 200 ff.

However, with the recovery of the Aristotelian corpus in the first century BC, new impulses are brought into philosophy. This primarily concerns Aristotle's doctrine of the intellect (νοῦς) and of God as self-thinking thought. The Aristotelian inspiration is met with in the *Didaskalikos* by Alcinous.[23] Alcinous teaches that there exists a supreme, immovable God.[24] This God is an intellect that eternally contemplates himself and his own thoughts. These thoughts or Ideas are the paradigm of the cosmos, but they are not the plan in the mind of the maker; because the maker, or the demiurgic intellect, is an intelligible reality 'below' the first God. God is somehow the cause of everything, but he does not act as efficient cause. The heavenly Mind or Intellect turns towards God and becomes an Intellect in act. In this way God orders the Intellect and the World-Soul in accordance with himself and his thoughts. The Intellect in act, probably receiving the Ideas in itself, creates an ordered world. As we see, before the emergence of Neoplatonism, Ideas, as divine thoughts, are conceived as contained in the divine intellect, but on a secondary level, due to the distinction between a first transcendent God and a 'lower' deity that relates to the material creation.

Neoplatonism originated with Plotinus (206–267/70). Compared with Middle Platonism it is characterized by a clearer doctrine of three levels of reality or hypostases, viz. the One, the Intellect (νοῦς), and the Soul. A very important feature is that the highest divinity, the One, is not conceived of as an intellect (νοῦς) as it normally is in Middle Platonism. In his perfect simplicity God transcends every predicate, every ontological category. It is not even adequate to designate him 'the One'.[25] According to Plotinus, everything has its origin in the supreme principle. By just being itself, and without being active as creator, the One is the source of the next hypostasis, the Intellect. In this creative process the *will* of the One plays an important role. I shall return to this topic of will when we come to the doctrine of creation.

[23] Dillon (1977, 268) follows convention and identifies Alcinous with Albinus. After careful consideration of the arguments of J. Whittaker, Dillon has eventually changed his opinion, and accepts that there are *no* good reasons to identify the two. Cf. his review in *Phoenix* 39(4) (1985), 420, of J. Whittaker's *Studies in Platonism and Patristic Thought*, Variorum Reprints, London 1984. Cf. Whittakers article '*Parisinus Graecus* 1962 and the Writings of Albinus' (Parts 1 and 2) in that volume. Cf. Alcinous, *The Handbook of Platonism*, ed. and trans. Dillon (Oxford 1995), ix ff.

[24] Dillon (1977), 280 ff.

[25] *Enn.* 5.5.6.

The One has an activity of the essence (ἐνέργεια τῆς οὐσίας) which is inevitably accompanied by an activity out of the essence (ἐνέργεια ἐκ τῆς οὐσίας). This activity culminates in the constitution of the hypostasis of the Intellect.[26] When I say that the activity *of* the essence is 'accompanied by' an activity *from* the essence, this does not mean that they are two completely separable activities. What occurs is an immanent activity with a transitive aspect. This is a general Plotinian principle.[27]

Now, how does he understand the hypostasis of the Intellect? Plotinus is concerned not so much with the question of the pattern according to which God made the world, as he is with the question of the relationship between eternal, intuitive thought and its object. Plotinus says that the Ideas (τὰ εἴδη) are contained within the Intellect, and it encompasses them as a genus does its species and a whole its parts.[28] But this can easily be misunderstood. There is no distinction between the subjective and objective side of this process of thinking. The act of thought and what is thought about are not two distinguishable moments in reality because the Intellect as a whole *is* all the Ideas. We are, therefore, to imagine an *identity* between the act of thought and its object.[29]

The Ideas as such are not primarily the paradigm of the lower world in the mind of its maker. But still the demiurgic function exists, and a paradigm of the lower parts of reality likewise. Surprisingly enough, the Intellect itself may still be seen as both creator and demiurge (ποιητής and δημιουργός), while 'nature', as some kind of basic subject, is the receiver of forms.[30] The cosmos has its origin in the Intellect, which therefore can be identified as the archetype of everything below it. The Intellect is an intelligible cosmos (κὸσμος νοητός), and this cosmos or archetype is identified with the paradigm of the Demiurge from *Timaeus*.[31]

In what way are we to conceive of all this? As I understand it, every level of being below the Intellect is in a *qualified* way the result of an intellectual activity. It is a general law of the Plotinian cosmos that as long as something remains in being, it is creative of a reality

[26] *Enn.* 5.4.2. [27] Cf. *Enn.* 6.1.20. [28] *Enn.* 5.9.6.

[29] *Enn.* 5.9.8. On the relationship between the Intellect and the Ideas, see Emilsson (1996).

[30] *Enn.* 5.9.3. [31] *Enn.* 5.9.9.

outside of itself. This reality is an image (εἰκών) of the being itself as archetype. The Intellect, therefore, is creative in the same way as the One, which is to say that it does not turn its attention 'anywhere' else, but remains with itself. This is its activity of essence. It creates (ποιεί) then, by manifesting a manifold power (i.e. the ἐνέργεια ἐκ τῆς οὐσίας). This 'power' or 'activity' (ἐνέργεια) manifested from the essence of the Intellect, is Soul. The Intellect itself remains unchanged through this creation.

The next stage in the process is the creation of the lower world of nature. Even here the Intellect has a causal role to play, but the demiurgic function is taken over by the Soul. The Intellect furnishes itself as paradigm, but is not in any way operative in the constitution of a material cosmos. Contemplating its origin in the Intellect, the Soul gets filled—but by what? We may ask. Probably by the Intellect as a paradigmatic intelligible cosmos. Turning towards itself (the ἐνέργεια τῆς οὐσίας), the Soul sets in order, directs, and rules (κοσμεῖ τε καὶ διοικεῖ καὶ ἄρχει, i.e. the ἐνέργεια ἐκ τῆς οὐσίας) 'what comes next to it' (the effect of the ἐνέργεια ἐκ τῆς οὐσίας). It moves and thus generates its own image (εἴδωλον), which is the immanent life-principle in plants and animals.

In what way does the Soul 'order, direct and rule' the cosmos? Plotinus compares the cosmos to a house that is presided over by its builder, who 'rules it while abiding above'. In a similar way, the Soul is present to the cosmos, without belonging to it. It masters the cosmos without being mastered by it. The cosmos extends as far as the Soul goes, and its being is constituted by a *logos* mediated by the Soul. This *logos* constitutes the universe from the ideal realm of the Intellect. It seems that according to Plotinus the *logos* somehow 'transports' the Ideas to the things that are created.[32] This 'transportation' by the *logos* is possibly a way of describing the activity of the hypostasis of the Soul in relation to the Intellect. The material cosmos, so to speak, comes from the intelligible cosmos of the Intellect, by the Intellect 'giving of itself' to matter. What the Intellect 'gives of itself' is the *logos*, which, as it is said, 'flows from it'.[33]

[32] *Enn*. 4.3.9. This is the opinion of Rist (1967) in the chapter about Logos.
[33] *Enn*. 3.2.2; cf. 3.2.16.

As we have seen already, Plotinus describes the relationship between Intellect and Ideas as the relation between a genus and its species.[34] I am not sure, however, whether Plotinus himself conceived of the Ideas as arranged in a complete 'Porphyrian tree'. Through the interpretations of Aristotle's doctrine of categories and predication given by Porphyry (234–301/5), the pupil of Plotinus, such a classificatory system becomes clear.[35] It remains uncertain who introduced this *detailed* classificatory system into the sphere of the Intellect. In any event the system is well established in Dexippus (beginning of the fourth century), a disciple of Iamblichus.

Dexippus wrote a commentary on the *Categories* in the genre of questions and answers.[36] In his treatment of *οὐσία* he takes Plotinus as his lead. Plotinus, Dexippus says, 'postulates one single genus of substance in the intelligible realm as being a common source of being to the incorporeal forms and thus bestowing being on the whole sensible realm and on the forms in matter'.[37] It seems probable that this 'one single genus of substance' is Plotinus' *νοῦς*, the second divinity in the Neoplatonic universe.

Natural objects have, according to Dexippus, an existence both in the intelligible and in the sensible sphere. This is an expression of that kind of Platonic realism which holds that, for instance, the human form in sensible beings is a participation in the perfect human Form in the unitary realm of Ideas. The perfect Ideas must be the intelligible version of those species and genera that are discoverable in the sensible world. We are able to discover the latter because we find an essential community among the things we experience. A species is discovered because it is held in common by the individuals, while a genus extends over the relevant species on a still more universal level. Beings are viewed 'from below', and by comparing them to each other one establishes the different levels of species and genera on the basis of the characteristics which join them together and separate one kind

[34] For Plotinus' rather complex account of the relationship between the act of thinking and the object of thought, cf. Emilsson: 'Cognition and its Object'.

[35] See Porphyry, *Isagoge* CAG 4: 4.9–5.6. We shall take a closer look at some sections of this text in Ch. 3.

[36] Translation in Dexippus, *On Aristotle Categories* (London 1990); Greek text in CAG 4.

[37] Dexippus, *On Arist. Cat.* 74; CAG 4: 40.28 ff.

from another. We recognize what is thus established as a Porphyrian tree. It seems to me that this abstract Porphyrian tree is the conceptual copy of the perfect, intelligible system contained in *οὐσία*, the *νοῦς* of Plotinus, as the primary source of being.[38] In this manner the Aristotelian species and genera, that is, the secondary substances of the *Categories*, were transformed into an actually existent plenary of intelligible Ideas in a principle which is the highest *οὐσία*.

We find the same line of thought in Ammonius (*c*.435/45–517/26), who brought the doctrines of Proclus from Athens to Alexandria.[39] Proclus himself had a doctrine of Ideas as exemplary causes for less perfect existences, but his teaching on these matters is so complex that it demands a study of its own.[40] In his commentary on the *Categories* Ammonius distinguishes between particular and universal substances.[41] He says that particular substances are not the basis for the existence of the universal ones, because the latter, which are the most honourable substances, preexist. When Aristotle says that particular substances are *primary*, he does not mean that they are primary *by nature*, but *to us*.

According to Ammonius, a distinction should be drawn between intelligible and sensible species and genera.[42] The immanent hierarchy is the forms in sensible things, and mirrors the transcendent hierarchy in the Mind (*νοῦς*). The intelligible species and genera, which are proper universal substances, are to be understood as the contents of the Mind, the demiurgic intellect. The Mind or the Demiurge has in its possession universal Ideas or *logoi* from the One, and through the Mind the One is both *causa efficiens* and *causa finalis* of the whole cosmos.[43]

This picture is highly reminiscent of the one we have already encountered in Dexippus, wherein the Porphyrian tree of abstract universals is transformed into a plethora of ideas that are contained in

[38] Cf. Dexippus, *On Arist. Cat.*, 60–1, 77–8, 82 ff., CAG 4: 29.12 ff., 42.13 ff, 45 (l. 12 ff.).

[39] Translation in Ammonius, *On Aristotle Categories*, London 1991, Greek text in CAG 4.

[40] Cf. Gersh (1978), 90 n. 43.

[41] Ammoius, *On Arist. Cat.*, 34 f., CAG 4: 25.5 ff.

[42] Ammonius, *On Arist. Cat.* 51, CAG 4: 40.23 ff.

[43] See Verrycken's article in Sorabji (1990).

a divine intellect. The epistemic logic of Aristotle is thus transformed into Neoplatonic metaphysics.

To end this section of developments within the Platonic tradition, some interesting aspects should be kept in mind as we move further into the Christian tradition and towards Maximus in particular. According to Antiochus the divine Ideas are conceived as dynamic *logoi* in God, while in Plutarch the Ideas are contained by the Logos of his system. Plotinus, on the other hand, distinguishes between what goes on in the Intellect and the *logoi* that convey the conceptions of the Intellect as an intelligible cosmos all the way down to the level of nature. Somehow Ideas proper and *logoi* are distinguished. However, both Antiochus and Plotinus conceive the *logoi* as rather dynamic principles. According to Dexippus there is an intelligible Porphyrian tree in the Intellect or Mind, and there is an immanent Porphyrian arrangment in the generated cosmos that can be abstraced from material conditions. Ammonius conceives universal substances or essences (*οὐσίαι*) as of a higher reality than the sensible ones. The *logoi* of Maximus are conceived as belonging to the Logos, they are quite dynamic in character and are related to a conception of a Porphyrian tree or a taxonomic arrangement. Still there are important differences. It should also be noted that according to Maximus universals are created,[44] and this shows us that his doctrine of *logoi*, as an exemplarist theory, have features that need a lot more elaboration.

IV. PHILO OF ALEXANDRIA AND EARLY CHRISTIAN EXEMPLARISM

With Neoplatonism there emerged cosmologies that culminated in the basic ideas of a transcendent divinity, an intelligible world situated in an intellectual hypostasis and of a demiurgic soul-substance. We now turn to thinkers of Jewish and Christian traditions. To the degree the Jewish philosopher Philo of Alexandria was studied by Christian theologians he could give them some clues how to interpret

[44] *Amb.* 7, PG 91: 1080a.

passages from the Bible. John Dillon considers Philo a Middle Platonic philosopher.[45] To Philo Moses was a great (Platonic) philosopher, and, says Philo, it is from Moses that the Greeks have their best thoughts.[46]

In *De opificio mundi* Philo develops his cosmology. He says that according to Moses, there are two basic principles, viz. the active cause (τὸ δραστήριον αἴτιον) and the passive element (τὸ παθητόν). The active principle is God, described as the Mind of the All (ὁ τῶν ὅλων νοῦς), transcending virtue, knowledge, the good itself, and the beautiful itself. God is, as we see, an intellect, but transcends every predicate. The passive element is in itself incapable of life and motion, but is set in motion, shaped, and quickened by the Mind, that is by God. The passive element is changed into a perfect masterpiece, that is the created cosmos.[47]

The created cosmos is a beautiful copy (μίμημα καλόν) of a beautiful paradigm (παράδειγμα καλόν). God gave form to the paradigm as an intelligible cosmos, a God-like and incorporeal pattern of the material world, before creating the visible cosmos. The corporeal world should contain as many sensible genera (αἰσθητὰ γένη) as the intelligible world contains intelligible genera (νοητὰ γένη).[48]

Philo's doctrine is similar to Plato's in some important respects. Philo made room for the paradigm of the Demiurge from Plato's *Timaeus*, a paradigm that contains the intelligible genera of sensible kinds. Even if Plato does not use the term 'intelligible cosmos', the doctrine of the paradigm as containing something like such a cosmos is present in the *Timaeus*.[49] There are, however, two important differences between Plato and Philo: (i) according to Philo the paradigm is created by God, and (ii) it exists in divine thought. But what divine thought is meant here? According to Philo the paradigm exists in the Logos. He constructs a parable, and says that when a king wants to build a city, a trained architect comes forward who forms the plan for

[45] Dillon (1977), ch. 3 C, 139 ff. [46] Dillon (1977), 140–1 and 143.

[47] *De op*. 7–9. What I have written above gives rise to the question of the eternity of matter. Philo is, it seems, ambiguous on this point of his doctrine, cf. Dillon (1977), 158. But in the *Legum allegoriae* (2.2) it is explicitly stated that 'before' creation, nothing was together with God. Cf. the interesting philosophical discussion of Philo by Sorabji (1983), 203 ff.

[48] *De op*. 16–17. [49] Cf. *Tim*. 30 c–d.

the city in his thought. The city formed in the mind of the architect has no place outside of him, but is engraved in the soul of the artificer as by a seal. Likewise, the cosmos of Ideas (ὁ ἐκ τῶν ἰδεῶν κόσμος) has no other 'place' than in the divine Logos.[50]

The Ideas exist in the Logos, but what or who is this Logos? Philo says more about this in the *Legum allegoriae*.[51] The Logos is described as the 'shadow' of God. It seems to be a kind of intermediary being, perhaps not a being with a complete hypostatic reality of its own, but a manifestation of God serving as His tool in the creation of the world.[52] The Logos, as an image (ἀπεικόνισμα) of God, is archetype for the creation of everything else. God Himself is the paradigm of the Logos, and the Logos is paradigm of all other beings. Through the activity of the Logos as divine tool, the Ideas accomplish their cosmological function by being implanted into creation as immanent *rationes seminales*.[53]

What then of the early Christians? Among the first Christian thinkers whose thought might be said to exhibit exemplaristic features are the Alexandrines Clement and Origen.

Clement of Alexandria thinks the Ideas are in the Mind (νοῦς) which is the Logos.[54] The Logos, Clement says, called Himself the Truth.[55] Now, Truth is an Idea, and an Idea is a conception which God has, and this Logos = Truth = Idea = conception which God has is the cause of creation. When Clement in another passage on the intelligible world implies that it is created on the first day of creation, this cannot mean that the Logos is created.[56] I would rather think he means that the eternal Logos of God the Father, as the Idea of Ideas, manifests within Himself the intelligible world in the creative act, when it is generated in matter at the first day of creation. The exemplarism of Clement comes to the fore when he says that the intelligible world is *archetypal*, and the sensible world is the image of that which is called the model.

50 *De op.* 17–20. 51 *Leg. all.* 3.96.

52 The scriptural background for the Philonian speculations of the Logos is to be sought in what is said about Wisdom in the Proverbs (ch. 8) and in the Wisdom of Solomon (ch. 7).

53 Cf. *Leg. all.* 3.150 and *De legatione ad Gaium* 55.

54 Clement, *Stromata* 4.25. 55 Ibid. 5.3. 56 Ibid. 5.14.

The exemplarism of Origen, together with his general cosmology, is very important in the history of Christian thought. Having come under suspicion of holding heretical views he was eventually condemned by ecclesiastical authorities in the reign of Justinian. The system of Origen is important in connection with St Maximus, because there are passages in the works of the latter directed towards a refutation of Origen and Origenist doctrines. There is, according to Polycarp Sherwood, reason to believe that Maximus had access not only to later Origenist works, but even to works by Origen himself, at least to *De principiis*.[57] St Maximus' aquaintance with the teachings of the great Alexandrian, whether directly through his writings or not, gave important impulses to the development of his own cosmological thought, not least to his theory of *logoi*.

According to *De principiis*, the beginningless Son of God the Father, called Wisdom, was 'begotten beyond the limits of any beginning that we can speak of or understand', and in 'this very subsistence of wisdom there was implicit every capacity and form of the creation that was to be'. Wisdom contained within herself the *initia*, *rationes*, and *species* of the whole creation.[58] It is reasonable to believe that the original Greek in this place read ἀρχαί, λόγοι, and εἴδη. In his commentary on the Gospel of St John, Origen likewise says that 'just as a house and a ship are built or devised according to the plans of the architect, the house and the ship having as their beginning the plans and thoughts in the craftsman, so all things have come to be according to the thoughts of what will be, which were prefigured by God in wisdom, "For he made all things in wisdom." ' In what immediately preceeds this, he explicitly locates the wisdom and plans of the cosmos to the Logos.[59] Now, the Greek term translated as 'thoughts' in this section is *logoi*. Another text from the same commentary shows that the sum total of *logoi* constitutes an *intelligible world* in the Logos, in accordance with which all things came into being.[60]

Both Philo and Clement teach that the intelligible world is somehow created, whereas the texts I have considered from Origen seem

[57] Cf. Sherwood (1955), 88 ff.

[58] *De princ.* 1.2.2, trans. Butterworth (1936), 16.

[59] *In Joan.* 1.22, PG 14: 56c–d, trans. R. E. Heine in Origen, *Commentary on the Gospel according to John, books 1–10*, Washington 1989, 57.

[60] *In Joan.* 19.5, PG 14: 568b–c, trans. in Wolfson (1976), 277–8.

to imply that this is *not* his view of the matter. But I would not dare to state this conclusion with too much confidence, because I am not intimately acquainted with his works, and much of what he has written has not survived. Clement's, and especially Origen's doctrine of the Logos containing all the *logoi*, gives the paradigm for later speculation over the divine Ideas in the Christian East.

It would be very useful to know whether there are exemplarist doctrines explicitly developed by influential Christian thinkers after Origin, especially among the Cappadocians. Exemplaristic thought in the Cappadocians would, I believe, have had a significant influence on later developments. But, unfortunately, as far as I can tell, there is not much to be found in the Cappadocians. We should first turn to St Basil of Caesarea. Jaroslav Pelikan quotes a passage from one of his letters and connects its meaning with the doctrine of divine Ideas: 'For the things of which we now see only the shadows, as in a mirror, later, when we have put on an indestructible and immortal body, we shall behold their archetypes (τὰ ἀρχέτυπα).'[61] I do not feel confident about connecting this quotation with a doctrine of Ideas. These archetypes or realities are, I suppose, identical with the ἐνέργειαι about which Basil writes in another of his letters.[62] Examples would be God's fearfulness, benevolence, justice, and creative power. 'Activities' such as these are understood by Basil to be somehow distinct from the divine essence. They are knowable, something which could not be said about the essence. The activities are some kind of divine manifestation by which God handles all His affairs with the created world. But all of this is too vague to be seen as expressive of an exemplaristic 'doctrine'.

In my opinion, we would do well to turn to Basil's *In hexaemeron*, because—if anywhere—we should expect to find exemplaristic teaching there. But once again the results are rather meagre, even though we do find some clues on the matter. In his second homily, Basil criticizes the doctrine of uncreated matter.[63] The proponents of such a doctrine are deceived by analogies from ordinary human life. Each art presupposes a plan and a material which is given form by a craftsman,

[61] Cf. Pelikan (1993), 98, cf. St Basil, *Letter* 8, 1: 89 (Loeb). Greek words inserted by me.

[62] St Basil, *Letter* 234, 3: 373 (Loeb).

[63] *In hex.* 2.2, PG 29: 29c ff.

and so one thinks that God must make for Himself a plan and work on a pre-existent matter as well. When Basil presents his own view he speaks of God, 'before' (πρίν) any of the visible objects existed, as 'having cast about in His mind and resolved to bring into being things that did not exist'.[64] This could imply that God Himself made a plan 'before'—in a logical, not a temporal sense—He created the world. God devised what sort of world He would create, and then He created the appropriate matter together with its form. This line of thought could be understood to indicate that Basil in reality has some kind of exemplaristic doctrine, even if it is not worked out in any details. But the results we may glean from what he actually says are, as I said above, rather meagre.[65]

If we turn to St Gregory of Nyssa, we get a clearer picture. In his *In hexaemeron* he tells us that God saw everything 'before' (πρό) their creation.[66] Everything was seen by the divine eye and was brought forth by 'the word of power' (ὁ τῆς δυνάμεως λόγος). This probably implies that God contemplated Ideas for everything that He wanted to create 'before' their actual creation.[67] The creation of the material world follows upon this in 'a regular order of development'.[68] God creates heaven, ether, star, fire, air, ocean, earth, living being, and plant, Gregory says. The thought of such a development is also found in Gregory's *De hominis opificio*.[69] Now, all of this, a divine plan or set of Ideas and an orderly development of creation, is indicative of a more securely grasped doctrine of exemplarism. Could this picture be confirmed by other texts of Gregory's? I believe it can. In his *De perfecta christiani forma* we are told that Christ is God's power and wisdom.[70] The whole creation, both what is perceivable through the senses and what transcends observation, was made by Him (δι'αὐτοῦ) and is united in Him (ἐν αὐτῷ).[71] The great wonders of creation would not exist if Wisdom had not thought of them, nor would

[64] St Basil, *Exegetic Homilies*, 24.

[65] In his treatment of the Cappadocian doctrine of Ideas, Pelikan (1993) does not provide any further documentation that would settle the matter in the case of St Basil. There is, of course, the possibility that somewhere in St Basil's works we could find clearer indications of his position, but so far such indications are unknown to me.

[66] *In hex.* PG 44: 72b.

[67] Cherniss (1934), 26, interprets the text this way.

[68] Ibid.

[69] *De hom. op.* 1, esp. 1.5, and 2. Cf. also 8.4-5. PG 44: 128c ff., 132d ff., 144d ff.

[70] *De perf.* PG 46: 260b. Cf. 1. Cor. 1: 24.

[71] Cf. Col. 1: 16–17.

they exist if Power had not accompanied Wisdom in the fulfilment of the divine thoughts (τὰ νοήματα). It is, Gregory says, through Power that the thoughts become deeds (δι' ἧς ἔργα τὰ νοήματα γίνεται). The divine Wisdom, Christ, had in His possession all the thoughts or concepts (νοήματα) 'before' the creation of the cosmos. These thoughts should reasonably be understood as divine Ideas for created things, Ideas that are to be realized in a created cosmos by Christ as Power.

In the *De virginitate* Gregory explicitly mentions the Idea of Beauty and the Nature of Beauty (ἡ τοῦ καλοῦ ἰδέα and τοῦ καλοῦ ἡ φύσις).[72] The text gives the impression that he is familiar with Plato's *Symposium*. The Idea of beauty is the cause (αἴτιον) of all beauty and all goodness, Gregory says.[73] This Idea of beauty, of course, may not be conceived of as a reality of its own, separated from God. It belongs to God, as one of His *attributes*. But if we think of it this way, then we have a problem. Divine *thoughts* seem to be one thing, divine *perfections* another. Which of these is entitled to the name 'Idea' when we speak about Ideas in connection with exemplarism? To me it seems reasonable to think that the texts from the *In hexaemeron* and the *De perfecta christiani forma* presuppose an exemplaristic doctrine of divine Ideas as divine thoughts or designs. The text from the *De virginitate*, however, is about something else. This may perhaps confuse the reader, so I have to make my point as clear as possible. What God saw 'before' creation, the things present in His mind, are Ideas properly speaking. These are the Ideas for those essences that God wants to create. But what Gregory terms 'the Idea of beauty' does not seem to be an Idea of an exemplaristic kind. It seems rather to be a divine perfection or a divine activity. We have to do with different aspects of the divine being. The first aspect is God's thought about what *kind* of things He wants to create. The second aspect is God's own perfections. Gregory has probably not systematically worked out this distinction. Later on I shall argue that the distinction between Idea and activity is established in Dionysius the Areopagite, and, to a much clearer degree, in St Maximus the Confessor.[74]

The third Cappadocian, St Gregory of Nazianzus 'the Theologian', gives clearer expression to an exemplaristic doctrine in his *Poemata*

[72] *De virg*. 11, PG 46: 364c. [73] Ibid. 368d. [74] Cf. Ch. 4 below.

Arcana. In the poem 'On the Universe', he first rejects the Platonic doctrine (cf. the *Timaeus*) that matter and form share the unoriginate status of God.[75] God is not like a painter who 'should produce a form resembling some other form, while observing before Him objects which His mind alone did not paint'.[76] Gregory teaches that God contemplated the Ideas of the whole economy within His mind, and then He created the intelligible and the sensible cosmos.[77] The same doctrine seems to be implied in Gregory's 'On Providence', but he is not as explicit in this poem.[78]

As we can see, there is at least a rudimentary form of exemplaristic doctrine in Cappadocian thought. Such a doctrine in, primarily, Gregory of Nyssa and Gregory the Theologian must have given a strong impetus to the development of exemplaristic doctrine in St Maximus. Dionysius the Areopagite is, of course, of special importance as well. We will return to him in the context of Chapter 3.

Since the doctrine of divine Ideas as the aspects of God's will is intimately connected with the doctrine of creation out of nothing, we shall now investigate Maximus' teaching on creation before we move on to details of his doctrine of divine *logoi* in Chapter 3.

V. ST MAXIMUS' DOCTRINE OF CREATION

'Exemplarism' is the doctrine that the world is created from divine Ideas. Exemplaristic doctrine, therefore, is intimately connected with the Christian doctrine of creation. Is the world created? Does it have a beginning? As will be seen below, it is possible to answer the first question affirmatively, but the second negatively. The doctrinal definitions from Nicaea (325) and Constantinople (381) in effect only answer the first question. But Christian philosophy and theology answered both questions in the affirmative, so that it became official Christian doctrine both that *the world is created*, and that *it has a beginning*. From the viewpoint of non-Christian thought Christian doctrine represented something unusual. Aristotle had maintained that the

[75] *Poemata Arcana*, 16. 3–4. [76] Ibid. 16. 21–3, trans. Sykes.
[77] Ibid. 20. 67–100. [78] Ibid. 22.

cosmos as a whole has neither beginning nor end. One can observe processes of generation and corruption within the cosmos, but absolute generation (γένεσις) or coming to be out of nothing, he ruled out, likewise for destruction into nothing. According to Aristotle, nothing comes from nothing. God, as described in the *Metaphysics* book Λ, is not the *causa efficiens* of the world, only its *causa finalis*. God does not concern himself with the welfare of the world or with human beings, but rests eternally in his own perfection.[79]

In Plato's *Timaeus* the cosmic order is made by the divine 'artisan', the Demiurge. The Demiurge brought a pre-existing matter from disorder (ἀταξία) to order (τάξις). The present cosmos, then, has originated as a living being endowed with soul owing to the activity of the Demiurge. He formed the cosmos after the pattern (παράδειγμα) of the Ideas.[80] The cosmos consequently, has a beginning, but not in the sense that matter has a beginning. In Neoplatonic thought in Late Antiquity the Demiurge looks like an Aristotelian divinity, while the highest principle is the One which is also called the Good. The Demiurge, contrary to Aristotle, is not only the *causa finalis* of the cosmos, but also its *causa efficiens*.[81] This way of thinking was quite common in Neoplatonic circles. One did not think that there were several worlds in succession, as did the Stoics. We might ask, then, whether they thought there was a beginning of the world, but not of matter, and whether God, as efficient cause, created order in pre-existent matter. It seems, rather, that the doctrine was different. If Plato is taken literally, the creation occured a set number of years ago. But this view of creation seems to be precluded from Plotinus onwards. The hierarchy of reality has no determinable temporal origin. There was never in the past any *time* when the levels in the cosmic order did not exist. If God (whether the first or second hypostasis) is the efficient cause of the cosmos, then creation must be *from eternity*. Even if this idea may initially look strange, it could be said to have some logical force. The Demiurge, i.e. the Intellect (νοῦς) of the Neoplatonic system, contains all the Ideas. The Intellect is unchangeable

[79] Cf. Aristotle, *Meteor.* book 1, ch. 14; *De gen. et corr.* book 2, ch. 10; *Physics* book 1, chs. 6–9.

[80] *Timaeus* 30a–c and 39d–40a.

[81] See e.g. Ammonius in Verrycken (1990). On Ammonius' doctrine of Ideas, see § III above.

and rests eternally in itself. Therefore, it seems reasonable to think that these Ideas from all eternity are mirrored in a material world.

The doctrine of creation was from the start a central Christian idea. In St Athanasius (*c.*296–373) and St Basil (*c.*330–379) the doctrine of creation received its classical formulation.[82] Athanasius attacks the Epicureans for teaching that 'the all' has come to be 'of its own accord and by chance' (*αὐτομάτως, καὶ ὡς ἔτυχε*).[83] Experience tells us that God both made and gave order to the world. Plato is spoken of with respect, but the theory of uncreated matter from the *Timaeus* is criticized. In this way, Athanasius says, the Platonists are imputing weakness to God, who could not have created anything unless matter existed. Against this Athanasius asserts that God is not just a craftsman (*τεχνίτης*) who fashions a given matter. He is a Creator (*κτίστης*) who creates the matter from which created things come into being.[84]

In his homilies on creation, Basil says that God created the world, heaven and earth. Created beings begin in time and end in time. Time originates with God's creation, from the beginning of the first movement. One may start with the present and reckon oneself backwards into the past and eventually discover the first day of creation. The world, which is finite and material, has not existed together with God from eternity. It is blasphemous to assert that matter is eternal in the same way as God. This would mean that matter merits the same kind of honour as God and should be venerated, Basil says.[85]

Now, it is one thing to establish what is orthodox Christian doctrine, but it is quite another matter to argue philosophically for its truth. According to R. Sorabji, up to 529 Christians adopted a defensive position in argumentation.[86] In 529, however, a member of the Alexandrian church, John Philoponus, who was a student at the Neoplatonic academy of Alexandria, moved to the attack. He tried to show that the cosmos *must* have a beginning. He published

[82] Among the earlier Fathers, St Justin is not at all clear on creation out of nothing, cf. *First apology* 10 and 59. In ch. 10 there is talk about creation ἐξ ἀμόρφου ὕλης. In Theophilus of Antioch there is, however, an explicit denial of creation out of uncreated matter, cf. *Ad Autolycum* 2.4 and 10 (Theophilus of Antioch, *Ad Autolycum*, ed. Grant, Oxford 1970).

[83] Cf. *De incarnatione* 2, in Thomson ed. (1971).

[84] Ibid.

[85] *In hexaemeron* 1.2–3, 6 and 7; 2.2, PG 29: 5c–12a, 16b–20c, 29c–33b.

[86] Cf. Sorabji (1983), 198, and (1987), 167.

a book called *De aeternitate mundi contra Proclum*, and later he followed up with a *De aeternitate mundi contra Aristotelem*.[87] The most important of his arguments have the Neoplatonist concept of infinity (τὸ ἄπειρον) in view. This concept is Aristotelian. Two arguments may be constructed: (i) If the world is eternal, then an actualized infinity emerges, because the world by now has existed infinitely many years. But this is impossible. (ii) If the world is eternal, then we will have to add to the infinite. If, for instance, an infinity of years are traversed by now, how many years will be traversed by next year? An infinity plus one year? But this is absurd.[88]

We must, then, according to Philoponus' arguments, accept that the world has a beginning. Sorabji comments on the results: 'My conclusion so far is that Philoponus' arguments are successful as an objection to Aristotle and the pagans. But the question remains whether we can answer his arguments by freeing ourselves in some way from Aristotelian ideas.'[89] We shall not follow Sorabji into his further discussion, but just note that according to Sorabji, with Philoponus a turning point occurs in the history of philosophy, because the Alexandrian for the first time puts Christianity on the offensive in the debate on whether the cosmos had a beginning.[90] Below we shall ask whether there are any traces of his arguments for the beginning of the world in St Maximus, who could well have read Philoponus.

When we now turn to Maximus we should expect the same picture as we found in Athanasius and Basil, and these expectations are confirmed. He considers the creative activity of God in several places, and I shall discuss some of the more important. He gives a short sketch of his doctrine in the opening chapters of *De char.* 4. St Maximus says that the world is created out of nothing, because of God's will and goodness, by His Wisdom and Logos. But what does it mean that the world is created out of nothing (ἐκ τοῦ μηδενός)? We shall see that it includes the teaching that the world has a beginning and consequently is not eternal.

87 The fragments of the latter work are collected and translated in Wildberg (1987).

88 Cf. Wildberg (1987), fragment 132.

89 Sorabji (1983), 217. Sorabji discusses possible objections to Philoponus' arguments from 217 on. He eventually disagrees with Philoponus.

90 Sorabji (1983), 224.

However, the doctrine of creation in Maximus poses some difficulties that should be considered further. These difficulties may be seen if we take *De char.* 4.3 as our startingpoint:

> Existing eternally as Creator, God creates when He wishes by His consubstantial Word and Spirit from His infinite goodness. But do not say: For what reason did He create now, since He always was good? Because, I say in turn, the inscrutable wisdom of the infinite essence is not subject to human knowledge.[91]

'Why did he create now, why not sooner?' This question, supposedly, originated in the fourth century, probably in Platonist circles. According to Sorabji, John Philoponus picks up the 'Why not sooner?' question from one of Proclus' arguments and tries to answer it.[92] Philoponus' opponent, Simplicius, criticizes Philoponus' answer and tries to reinstate the problem.[93] We have seen that the Neoplatonic Demiurge or Mind contained all the Ideas for the created world. It seems reasonable to think that these Ideas in all eternity are mirrored in a material cosmos. Maximus says that God is 'existing eternally as Creator'. He further says that God manifested His 'eternally pre-existing knowledge of beings'.[94] This 'knowledge' is the *logoi*, and these *logoi* are divine wills. If thought and will are identified in God, then, from a philosophical point of view, it should be reasonable to suppose that what God thinks, He wills, and He wills it from all eternity. There should, then, be no reason for God not to create sooner, rather He should be expected to create *infinitely* sooner, i.e. from eternity. Consequently God has created the cosmos from eternity.[95] Even though Maximus' premises take him close to a Neoplatonic position, he does not draw this conclusion. Why not, and how will he defend his position?

In the quotation from *De char.* 4 above, Maximus plainly rejects the question about *why* God created the world with a temporal beginning in the past. We shall now see what a further analysis of the context reveals. *De char.* 4.1–13 is a thematic unity.

[91] PG 90: 1048c. [92] Sorabji (1983), 236. [93] Ibid. 232.

[94] *De char.* 4.4, PG 90: 1048d.

[95] According to St Athanasius, even if God is Creator from all eternity, this does not mean that created being is eternal. He focuses on God's will and His freedom to create. Cf. *Orationes contra Arianos* 1.8.29, PG 26: 72a ff.

At the beginning of the text, Maximus says, apophatically and axiomatically, that when one is amazed at how everything is brought to being from nothing, one must remember that divine greatness has no end,[96] and that God's prudence (φρόνησις) is inscrutable (4.1, PG 90: 1048b). When it comes to God's creative activity, Maximus speaks against 'the Greeks', probably against Neoplatonic theories about the eternity of matter or the world. Polemical questions like 'Why now? Why not sooner?' are rejected with the words 'the inscrutable wisdom of the infinite essence is not subject to human knowledge'. One may, Maximus says, be filled with awe when one reflects on 'how and whence' (πῶς καὶ πόθεν) creatures originated when there was no matter prior to their existence; but one should not ask *why* God created now.[97] Man may even seek knowledge of the reason (ἡ αἰτία) why God created the world, but one should not try to understand 'how and why' (πῶς καὶ διὰ τί) He created recently (προσφάτως), since it is impossible for this subject to be grasped by the human mind.[98] Maximus emphasizes the following apophatic principle: τὰ περὶ αὐτό may be known, but τὰ κατ' αὐτό are not knowable.[99] The first kind of elements are manifestations of the divine being *ad extra*. They are the divine activities by which He creates the world. The second kind of elements regard the divine being itself, how God is in Himself.[100]

The outcome of this is that man may legitimately philosophize about the relation between God and the world. He may ask the question of *how* creatures originated, and he may ponder the question of *why* God created. The primary data of these subjects are somehow revealed in nature and Scripture, even if mysteriously. But why God created *recently* is not knowable because it pertains to the divine being itself and knowledge of the divine being transcends the capacity of every created intelligence. From a human perspective, the question is unanswerable. On the one hand, if viewed from Maximian principles, the question 'Why not sooner?' could even be considered meaningless because there is no time 'before' the creation of the world, and therefore nothing which could be 'sooner'. This, in effect, is St Augustine's

[96] Psalm 144: 3 (LXX); 145: 3.
[97] *De char.* 4.2–3, PG 90: 1048b–c.
[98] *De char.* 4.5, PG 90: 1048d.
[99] *De char.* 4.7, PG 90: 1049a–b.
[100] *De char.* 1. 100, PG 90: 982d–984a.

answer.[101] One could add that within the horizon of time it makes sense to tell why something did not occur earlier. There could even be given sufficient reasons: Socrates could not have been born that much sooner, because his parents had not yet met.[102] Concerning the creation of the world, however, one could not point to any previously given reasons, because God is beyond time and no relevant conditions can be identified. If the point of the question is why the world is not *older* than it actually is, then it could be taken to ask why the historical acts of the divine economy are fulfilled at just this and that point of time. The Maximian answer to this challenge is that it is a divine mystery, connected with divine providence. The fullness of times (πλήρωμα τῶν καιρῶν), talked of by St Paul (*Eph.* 1,10), is known by God in the secret depths of His being.

Maximus' teaching on the recent creation of the world shows that he firmly believes it to have a beginning. Two further comments of his in *De char.* 4, makes this clear beyond doubt: while 'the Greeks' teach that God did not create the substance of things, but only the properties, the Christians teach that God created the substance endowed with properties. This could only mean that neither realized creatures nor matter exist from eternity.[103] The *problem* of the eternality of the world, however, does not seem to be rejected. He actually argues against the idea that the world is eternal.[104] In his argument he stresses the impossibility that created things could exist from eternity together with God. What is limited in every way cannot exist forever with what is wholly infinite, he says. Creatures are not to be called creatures if they have eternal coexistence with God. But is this good enough to reject the Neoplatonic doctrine? Why is coeternity to be rejected?

We shall investigate some of the texts in which Maximus deals with this question. But before we dive into the intricacies of his arguments, we shall ponder briefly why he takes care to argue against pagan Neoplatonism in these matters. The heyday of pagan Neoplatonism had come to an end before Maximus was born (*c.*580). Why on earth should he bother arguing against a rival school of thought which

[101] *De civitate Dei* 11.5–6. [102] Sorabji (1983), 232.
[103] *De char.* 4.6 and 2, PG 90: 1049a and 1048b–c.
[104] *De char.* 4.6, PG 90: 1049a.

had ceased to have an influence on men's minds? There are some reasons why he should bother. First, one should note that Christian metaphysics itself had for a long time developed in critical tension with Neoplatonic thought. Christians and Neoplatonists to a large degree shared a common vocabulary if not a common set of concepts. The Christians, of course, exploited the vocabulary in order to formulate their Christian conception of the cosmos. But the intellectual affinities which existed between the two 'systems', affinities which would last even when pagan Neoplatonism had lost its force, made it a challenge for later writers in the Eastern Christian tradition to draw dividing lines between Christian ideas and Neoplatonic ones on crucial matters. Second, one could not be sure that pagan Neoplatonism could not revive as long as its written remains still existed and were studied in Christian schools. The history of Byzantium shows that such revivals were possible. The closeness of the Byzantines to the classical heritage created a continuous challenge. On this background one could explain a certain feature of Maximus' arguments. They do not seem to be developed from a close examination of Neoplatonic texts, rather he seems to argue against Neoplatonic positions that are constructed to be the typical targets of Christian criticism. One should keep this in mind when working with his arguments so that one does not take Maximus' characterization of his philosophical opponents to be an exact description of their views. There is no reason to believe that Maximus himself has consciously distorted Neoplatonic doctrine, rather he considers what is traditionally held among Christians to be the typical Neoplatonic positions.

The three texts of Maximus that we shall look into are all found in the *Ambigua*. We shall first turn to the seventh *Ambiguum*, where he is arguing against Origenism, especially against the idea of the pre-existence of souls.[105] But as the argument proceeds, what he in fact works out becomes a metaphysical alternative to Origenist and Neoplatonist metaphysics. Maximus says that God knows all things before their coming to be.[106] They were, however, not brought into being together with the divine knowledge of them, 'but each of them gets actual being at the proper time in accordance with the wisdom of the Creator' (ἀλλ' ἕκαστα τῷ ἐπιτηδείῳ καιρῷ κατὰ τὴν

[105] Cf. Sherwood (1955), 21 ff. [106] *Amb*. 7, PG 91: 1081a–b.

τοῦ Δημιουργοῦ σοφίαν ... εἶναι τῇ ἐνεργείᾳ λαμβάνῃ). The Creator Himself is always actualized, while creatures before creation represent a mere possibility. They are possible qua known by God. Maximus denies the possibility of God and creatures to have being simultaneously (*ἅμα*). What is unlimited and what is limited, the superessential[107] and the essential cannot exist in metaphysical 'simultaneity' or have being 'together'. What is measurable and what is immeasurable, what is conditioned and what is unconditioned, what is determined by all categories and what cannot be determined by any of them, may not be conducted into the same (*εἰς αὐτὸν ἀγαγεῖν*), Maximus says.

In the above, Maximus formulates some important points: even though beings are *known* eternally (in their *logoi*), they do not *exist* eternally. It is one thing to be known by God, quite another thing to exist. What is known by God in eternity is given being by Him 'at the proper time' (*τῷ ἐπιτηδείῳ καιρῷ*). As known by God, beings are potential with a view to real existence. According to this line of thought, then, beings have a beginning of existence from the proper moment, 'in accordance with the wisdom of the Creator'. These last words should be understood as indicating a reference to the role of the divine will. God's *will* to create is stressed at least three times in the opening chapters of *De char.* 4: Maximus says that God creates 'when He wishes' (*ὅτε βούλεται*), that He produced beings 'when He willed' (*ὅτε ἐβουλήθη*), and once again that He can give substance to something 'when He wishes' (*ὅτε βούλεται*).[108] We can conclude, then, that in Maximus' view, God created the world, giving it a beginning of duration, in such a way that it came to be in accordance with what He willed in His wisdom.

This line of thought, however, is not an *argument* for the beginning of creatures. It is a *statement* of how such a beginning is to be understood. The next turn of Maximus' reasoning is the denial of simultaneous existence of God and creatures. However, it may be that a beginning of temporal duration of beings is not in the end established along these lines either.

[107] The 'super' shall be taken in the sense that God transcends essence. He is 'transessential'.

[108] *De char.* 4.3-4, PG 90: 1048c–d.

Maximus probably considers his denial to be intuitively evident. If God and beings should exist in metaphysical 'simultaneity', they would, according to his view, have to share identical modes of being. This implies that creation or matter would reach a divine status, something that is unacceptable according to Christian beliefs. There is an implicit premise in Maximus' argument that is expressed in St Basil's *In hexaemeron*.[109] Basil says that the whole of anything, whose parts are subject to corruption and change, must also at some point submit to the same changes as its parts. This idea, according to the report of Diogenes Laertius, is a Stoic doctrine:[110] 'And that of which the parts are perishable is perishable as a whole. Now the parts of the world are perishable, seeing that they are transformed one into the other. Therefore the world itself is doomed to perish.' Basil and Maximus seem to claim that if created things existed together with God from all eternity, they would have divine attributes, such as incorruptibility, immutability, and superessentiality. But experience shows that this is not the case with the parts of the created world, consequently it cannot be the case with the whole. Therefore, since creation is corruptible, mutable, and essentially determined, it cannot exist with God on the same level of reality as eternity. Since there are different sets of predicates, there must be different substances. One of these, viz. creation, has its being from the other, and the ontological status of created being must be totally determined from its cause as something 'outside of' the cause and something completely 'other' than it.

Is this way of thinking logically sound? The attributes of beings may be determined from experience. The parts of creation of which we have experience are corruptible and mutable, etc., but is it therefore correct to think that the totality too has these attributes? If the totality is just the sum of the parts, then it probably will have the attributes of the parts. The parts we actually experience are experienced as corruptible, etc. But are we entitled to draw a conclusion from *some* to *all*? Must we not experience *all* the parts to draw conclusions about their totality? We have here a problem of induction. But even if that is so, is it not the case that even if we do not examine

[109] *In hex.* 1. 3, PG 29: 9a–12a.

[110] Diogenes Laertius 7.141. Hicks' translation (Loeb).

each item in the totality, we will still be able to recognize that mutable parts exist in it and somehow set their stamp on it? It cannot, then, if that was claimed, be incorruptible, immutable, etc. in all respects. Consequently, it would not have such attributes in the same way that God has His. In several ways there would be *contrary* predicates, and it seems reasonable to think that God and creation could not have their being with metaphysical simultaneity. To Maximus it is an obvious fact that they cannot exist together, rather they must belong to two different ontological spheres. The ontological 'otherness' of God is emphasized by Maximus both in *De char.* 4 and in *Ambiguum* 7.[111]

To the degree that Maximus considers the totality as something more than the sum of parts, this 'more' is not an ordinary element of the totality. The 'more' will be the unifying and ruling principle that comes in addition to the totality. In the first chapter of the *Mystagogia*, Maximus distinguishes between the parts, the whole (ἡ ὁλότης) and the principle of the whole.[112] The parts are created beings. What I have termed the 'totality' of such beings is by St Maximus considered a 'whole'. This 'whole' is constituted by an orderly arrangement imposed by God on the parts. As such it is not an element resulting from the gathering of the parts, but is instituted by God as the hierarchical order of creatures. Only in light of this whole are the parts adequately understood, because the whole reveals the parts in the divine scheme of things, in the *oikonomia* of the created cosmos. The principle of this whole (i.e. the orderly arrangement of the cosmos) is the *logoi of beings*. The one Logos of God the Father gathers this totality together into unity. The collection is illustrated by the image of the circle, its centre and radii, in the last part of the first chapter of the *Mystagogia*. The whole arrangement of beings and the basic principle of this arrangement do not exist under the same ontological conditions. The principle, which is God comprising the collection of divine Ideas, *transcends* the world, but is at the same time somehow at work *immanently* in it. To Maximus it is important to preserve both the divine transcendence and immanence.

What does the argument from *Ambiguum* 7 accomplish? Strictly taken it does not establish a temporal beginning, but only the

[111] Cf. *De char.* 4.7–9, PG 90: 1049a–b and *Amb.* 7, PG 91: 1081b.
[112] PG 91: 664d ff.

ontological priority of God over created beings. The argument differs from, but might be compared with, an argument in Proclus about the priority of the first principle.[113] Proclus, as may be expexted, emphasizes that his argument does not establish a temporal beginning of 'otherness'. This, of course, is perfectly in his interest. Maximus, I think, is no better off, even if he would have preferred to be so. On the other hand, his argument is not at all incompatible with a doctrine of a beginning. What he has demonstrated with his argument and his belief that the world has a beginning is consistent, even if the doctrine of the beginning of the world is not *philosophically* demonstrated. But did he really consider it demonstrated? I really doubt it.

The next text we shall consider is from *Ambiguum* 10.[114] Maximus is constructing an argument from *motion* for the beginning of creatures. First he says that contemplating the beauty and greatness of God's creatures, one understands God as beginning, cause and maker (ἀρχή, αἰτία, and ποιητής) of them. Later on he talks about 'the error that the world is ἄναρχος', that is without beginning or origin (ἀρχή). Does the word *arché* have the same sense in these two expressions? If the argument he constructs is against Neoplatonic thought, which is highly plausible, then we should remember that the Neoplatonists would have argued that God is the *arché* of the world, and that the world for that reason is *not* without beginning (ἄναρχος), i.e. without an ontological source. What they would have denied is that the world had a beginning for its temporal existence. The error, to which Maximus is referring, is the Neoplatonist denial of a *temporal* beginning. For this reason I believe that *archē* is here understood in two different senses by Maximus, the two senses being (i) ontological origin of being (God is the *archē* of the cosmos), and (ii) beginning of temporal duration (cf. the cosmos is 'not ἄναρχος').

It is an error, then, to deny a temporal beginning. Why? Maximus next presents his argument from motion, which is not set forward in all details in the text, but may be reconstructed thus:[115] everything that is in motion has a beginning, because everything that is in motion has a cause, and everything that has a cause has a beginning.

113 Proclus, *The Elements of Theology*, prop. 5.

114 *Amb.* 10, PG 91: 1176d–1177b.

115 Could these arguments from motion stem from John Philoponus? Cf. Philoponus ed. Wildberg (1987), book 6, esp. fr. 109.

A further argument could be constructed from this, which, however, does not turn up in the text: the world has a beginning, because the world is in motion, and everything that is in motion has a beginning. Again, 'beginning' here means 'temporal beginning'. We can present these two arguments as two syllogisms that are chained together. The conclusion of the first is a premise of the second:

P1	Everything that has a cause has a beginning.
P2	Everything that is in motion has a cause.
C1/P1′	Everything that is in motion has a beginning.
P2′	The world is in motion.
C2	The world has a beginning.

Even if the second argument is not put forward in the text, it establishes the truth that Maximus eventually wants us to acknowledge. The arguments are valid, but their truth depends on the truth of the premises. The P2 of the first syllogism seems plausible, but what about P1? The Neoplatonists would have denied it if 'beginning' is to be understood in the temporal sense. I believe that they willingly would have conceded the whole argument had the word *archē* been understood generally as 'origin of being', but this would destroy the Maximian point. Maximus would have said that P1 is not meant theologically or ontologically, but temporally. Do there remain any possibilities for St Maximus to argue his first premise as a premise about temporal beginning? In the text it seems that he is aware of the problem, and he tries to make his point about a temporal sense when he claims that an effect (i.e. 'something which has a cause') has two *termini*, a *terminus a quo* and a *terminus ad quem*. An efficient cause and a final cause exist for every motion, and the movement between these is reasonably seen as a movement in time. If anything is moved, then it is moved temporally. If something has an efficient cause and is moved temporally towards its end, then its motion has a temporal beginning. The Neoplatonists, I believe, would have conceded all of this, except that there must be a temporal beginning for temporal motion for the cosmos as a *whole*. The only way Maximus could have defended his argument by now would have been by attacking the Neoplatonist concept of infinity in the way that John Philoponus did. This would have saved the whole argument. Are there any traces of Philoponus' way of argumentation in this context of the *Ambigua*? I shall consider one additional text.

In a long section from the tenth *Ambiguum* Maximus discusses the concept of infinity.[116] The theoretical content of this section is highly complex, but while I may not have understood it correctly in all details, that should not create any difficulties for our present purpose, since the main line of argument can be grasped, and it is this main line which concerns us. Interestingly enough, the concept of infinity that Maximus dwells on here is quite different from the one that John Philoponus uses in his arguments against the Neoplatonists. Philoponus took his point of departure from a concept of infinity that made it possible for him to construct an argument against those who taught the eternality of the world. Maximus develops a concept that, according to his beliefs, *positively* excludes the idea of an eternal matter or an eternal world. Infinity in Maximus' argument is something totally realized. He thinks a totally realized infinity could never be material, because it has properties which matter does not have.

The main argument[117] is introduced with the following reasoning:[118] if matter existed absolutely,[119] then it did not come into being. If it did not come into being, then it was not moved. To turn it around: if it was not moved, then it did not begin to be. If it did not begin to be, then it is totally beginningless (ἄναρχον). If it is beginningless, then it is infinite. If it is infinite, then it is unmoved. Maximus adds the comment that the infinite is unmoved because as infinite it has no place in which to be moved. The conclusion is that if this is the case, then there must be two infinities, both God and matter, with the same sets of properties. But this, he says, is unthinkable.

116 *Amb.* 10, PG 91: 1181a–1188c.

117 There is an argument before the main one is introduced. But it seems to ascribe to the Neoplatonists—if, as I believe, it is directed against them—a doctrine which I do not think they would have accepted, viz. that matter possesses being by itself and not as received from God. Cf. *Amb.* 10, PG 91: 1181c–1184a.

118 *Amb.* 10, PG 91: 1184b ff.

119 *Amb.* 10, PG 91: 1184b. The Greek has εἰ ἦν . . . ἡ ὕλη, which Louth (1996), 141 translates 'if matter was [absolutely]', cf. his note 101 (209). I agree with Louth that this is what is meant.

Why is it unthinkable? Maximus could by now have appealed to experience that clearly shows movement taking place, or he could have argued against the conception of two infinities. What he does, however, is to argue that according to the way matter is understood, it does not have the properties of infinity. In the argument matter is spoken of as 'the dyad' (ἡ δυάς) which, according to Aristotle, is the second one of the two highest principles in Plato, the first being the One.[120] These two principles are in the Platonic system the origin both of the Idea-numbers and in the end of sensible things. Plato does not speak of the dyad in his dialogues, but he might well have used the expression, for instance in his lecture 'On the Good'.[121] The term is used in Middle Platonic and Neoplatonic speculation over principles, and in the commentators on Aristotle, for instance in Alexander of Aphrodisias.[122] From this history we cannot, however, read anything into the sense of the term as used by Maximus.

The infinite is, Maximus says, neither divisible nor divided, neither composite nor compound, neither is it division nor composition itself, it is neither numerable nor numbered nor co-numbered, it cannot enter into any relationship at all.[123] The dyad, on the other had, could not possibly be qualified thus, the contrary is rather the case.[124] Therefore, according to Maximus, the dyad is not infinite, not without beginning, not without motion. If this line of argument is sound, one could claim that the material principle, 'the dyad',

[120] Ritter (1972), 302. Cf. Aristotle, *Metaphysics* A, 6. 987^{b} 18 ff., cf. Plato, *Philebus* 23c–26d, Arist. *Physics* book 3, ch. 4. 203^{a} 15–16.

[121] Peters (1967), 42.

[122] Ibid. and Ritter (1972). Alexander of Aphrodisias, *On Aristotle Metaphysics 1*, trans. by W. E. Dooley, London 1989, 84 ff. (CAG 1: 55.20 ff.), *On Aristotle Metaphysics 2 and 3*, trans. by W. E. Dooley and A. Madigan, London 1992, 140 and 182 f. (CAG 1: 203.25 ff., 228.1 ff.).

[123] Some of the properties which are denied of the infinite, I find hard to understand. Maximus seems to deny that the infinite is 'sole and simple' (ἁπλοῦν καὶ μόνον), cf. the translation of Louth (1996), 141. Could it be that what Maximus wants to say is that the infinite is neither division nor composition, because *divison and composition* are neither sole nor simple? It is further said that the infinite is not 'simply free from any kind of relationship'. Could it be taken to mean that the infinite is not 'as it happens' (i.e. acidentally) free from relationship, because it is *essentially* or *necessarily* transcending relation?

[124] Cf. Alexander of Aphrodisias trans. Dooley (1989), 85 (CAG 1: 56.10 ff.), where the dyad could be understood to have several of the qualities contrary to Maximus' infinite.

has a temporal beginning. But why the cosmos was created *recently*, remains a divine mystery.

Whether or not Maximus succeeds with his *arguments*, his *doctrine* is clear in all essentials: beings are created by God *out of nothing*. This means that creation, whether matter or the completed order, is *not eternal*. Creation (matter or cosmos) has *a temporal beginning* so and so many years ago.

According to Maximus, creation was brought from non-being to being by God.[125] While the divine essence (*οὐσία*) does not have a contrary, the essence of beings has its contrary in non-being (*τὸ μὴ ὄν*).[126] Beings are kept in being by absolute dependence on the power of God. Separated from Him their being will dissolve into non-being.[127] Consequently, the totality of beings is a work of *creation* in the proper sense.

The ontological status of created being is totally conditioned by its cause. The knowledge of being existed from eternity in God, and He instituted (*ὑποστήσας*[128]) its *genesis* before the ages by His will alone, Maximus says.[129] This does not mean, as we have seen by now, that the world is eternal. Nor does it mean that God made His decision infinitely long *before* the creation nor that He at *a given moment* made it. Both ways of expression place God under the category of time. It is almost impossible to formulate this adequately because all human thought is determined by time. However, it is after all tempting to try, and perhaps we could say something like this: the eternal God has eternally willed that something 'other' than Himself should originate with a beginning of its temporal existence.

For Maximus, creation is not due to necessity. God is a free Creator. This is strongly emphasized, as we have seen, in *De charitate* 4. The emphasis placed on God's creative will is a characteristic of Christian thought, and is usually considered to be a dividing line between Christianity and Neoplatonism. It is often said that Neoplatonic thought is marked by what is called an 'emanationist' doctrine.

[125] Cf. *De char.* 4.1, PG 90: 1048b; *Amb.* 7, PG 91: 1077c.

[126] *De char.* 3.28, PG 90: 1025b–c.

[127] *Myst.* 1, PG 91: 668b.

[128] The verb *ὑφίστημι* means, for instance, 'give substance to', 'cause to subsist', 'institute'.

[129] *Ad Thal.* 22, CCSG 7: 137. 4 ff.: *'Ο πάσης κτίσεως, ὁρατῆς τε καὶ ἀοράτου, κατὰ μόνην τοῦ θελήματος τὴν ῥοπὴν ὑποστήσας τὴν γένεσιν πρὸ πάντων τῶν αἰώνων.*

Plotinus says the great multiplicity flowed from (ἐξερρύη) the One, and that it somehow overflows (ὑπερερρύη), and 'its superabundance makes something other than itself'.[130]

On this background one could assert that, according to Plotinus, everything that comes 'after' the One, is generated by flowing out from the One by some kind of necessity. It would, however, be wrong to take this metaphor of 'flowing' (*emanatio*) as an expression of exact doctrine. Plotinus says that the One acts according to its *will* (βούλησις).[131] The case with Christianity is obviously not solved yet. The question is, however, what character this will has. To find an answer we are directed to another text from Plotinus:

> All things which exist, as long as they remain in being (μένει), necessarily produce from their own substances (ἐκ τῆς αὐτῶν οὐσίας), in dependence on their present power, a surrounding reality directed to what is outside of them, a kind of image of the archetypes from which it was produced: fire produces the heat which comes from it; snow does not only keep its cold inside it.'[132]

This passage too is filled with metaphor, but it contains two important points: (i) as long as a being 'remains' (μένει) it produces, and (ii) it produces necessarily. The One is not subject to any *external* constraint, and does not act by necessity for that matter. But could there be an *internal* constraint, which determines that the One acts the way it does? The clue to answer this question lies in the idea of 'remaining'. According to John Rist, the question which has to be answered is 'What is the One that it emanates Νοῦς and Being?'[133] Rist shows that the solution has to be sought in the concept of *activity of the essence* (ἐνέργεια τῆς οὐσίας) with its aspect of *activity out of the essence* (ἐνέργεια ἐκ τῆς οὐσίας). With this concept we move from the sphere of metaphor to the sphere of philosophical doctrine. As far as I can understand, the point is that if the internal activity of the One *remains* in a certain way, then the external effect of this activity *necessarily* is in a certain way too.[134] Now, why should the internal activity of the One remain in a certain way? The answer is because

130 *Enn.* 5.1.6 and 5.2.1.
131 *Enn.* 6.8.13.
132 *Enn.* 5.1.6. Armstrong's translation. Greek terms inserted by me.
133 Rist (1967), 69.
134 Ibid. 70 to the end of the chapter.

the One wills to be itself.[135] The whole point is that in willing to be itself and therefore in remaining itself, the One naturally creates something 'outside' of itself. Is it reasonable to call this will to be oneself an internal constraint? Could it not rather be called freedom? This is a difficult question, but I believe that within certain systems of thought it is reasonable to understand freedom in this way. One could of course ask the impertinent question of whether the One could have willed not to be itself. If this possibility does not occur, one could claim that the One is not free after all. But according to Plotinian and, I believe, Christian understanding, this would amount to a misunderstanding of divine freedom. The One is the Good, and to will something other than itself would mean to will what is contrary to the divine nature. On the divine level, to be free means to live according to the divine nature.

The One of Plotinus is also the Good. When the One wills to be itself, it wills itself as Good. Willing itself as Good, can be described as the activity of the essence with its natural concomitant the activity out of the essence. This could mean that the One wills itself as diffusive of good. Here we should remember that the reason that motivated the Demiurge of the *Timaeus* to create was his goodness.[136] Now, all of this is, at least apparently, not different from what a Christian understanding of the Creator and His work of creation would be. For Maximus, the reason why God created the world was His infinite goodness.[137] The divine goodness is then, as in Plotinus, diffusive of itself. This goodness cannot be separated from God's eternal knowledge of beings.[138] It would not be wrong to say that God, willing Himself, wills Himself as good and as knowing what He wants to create. Inevitably the result is, in Plotinus and in Maximus, that God creates something 'other' than Himself.

However, there are differences between the Neoplatonist and the Christian view. One such difference concerns the beginning of created being. In Plotinus created being does not begin its existence so and so many years ago. On the Christian side, St Athanasius said that even if God is Creator from eternity, He did not create the world from eternity.[139] As we have seen above, this is exactly what Maximus

[135] *Enn.* 6.8.13. [136] Cf. § II. [137] *De char.* 4.3, PG 90: 1048c.
[138] *De char.* 4.4, PG 90: 1048d.
[139] *Orationes contra Arianos* 1.29, PG 26: 72a ff.

asserts as well. This points to a difference between the Christians and the Neoplatonists in their understanding of the divine *will*. According to Christian beliefs, God is free to will that the world should begin to exist as a limited reality with time. As we have seen in Maximus, this doctrine is difficult to justify philosophically. We should not, however, forget the partial succcess of John Philoponus' arguments against the Neoplatonic doctrine of an eternal world. We have to admit, though, that the Philoponian reasoning is from 'below', i.e. from the created world. To reason from 'above', i.e. from the point of view of the divine being, on the other hand, is to establish arguments from a reality that is beyond the capacity of the human mind.

To move on a bit further, we could ask Maximus whether God could have willed that the world should *not* exist after all, and still remained in His self-identical goodness by nature. This is a difficult question to answer. One should, maybe, think that a Christian would have to answer in the affirmative, but Maximus claims that God 'always and in all' (ἀεὶ καὶ ἐν πᾶσιν) has a will to 'embody' Himself in created being.[140] Here, however, the question about the sense of this 'always and in all' turns up. There is no doubt that the divine 'embodiment' is not to be taken just in the sense of the historical Incarnation of Jesus Christ. The 'in all' has cosmological implications as well.[141] When Maximus says that God 'always' has this will to embody Himself, it means that God willed His embodiment from eternity. Even the historical Incarnation, according to Maximus, has its origin in the super-infinite plan that infinitely pre-exists the ages of time.[142] This shows that the total economy of creation and Incarnation belongs to God's *eternal* plan, so that God, because of His natural goodness, would never have refrained from creating the world. God, remaining Himself with His eternally good intention, created the world and gave it a beginning of its existence. I suppose we have to say that in His freedom God remained faithful to Himself and excluded the possibility of not creating the world.

This, however, even if it partially answers the initial question, poses a fresh problem. If in creating the world God 'remained faithful to Himself', one could ask whether God, if deciding not to create, would

[140] *Amb.* 7, PG 91: 1084c–d. [141] *Amb.* 33, PG 91: 1285c–1288a.
[142] *Ad Thal.* 60, CCSG 22: 75.40 ff.

not be faithful to Himself and therefore would have suffered a change in His eternal being. If He actually would have suffered such a change, God, obviously, *could not* refrain from creating, and then we must admit that there exists some kind of *necessary relation* between God's nature and creatures. Even if the world has a temporal beginning, God by necessity had to relate Himself to creatures by becoming their Creator. According to the *Cap. gnost.*, however, God infinitely transcends any relation to creatures.[143] How could this difficulty be solved?

Maximus has committed himself to the basic idea that the divine economy is anchored in the inner life of the Trinity. I believe that the question of whether God would have suffered any change had He refrained from His creative activity would have been meaningless to Maximus. The point is that God from eternity has freely *chosen*, not only to know the mystery, but even to realize it in actual fact. The created world is totally dependent upon God's will and power, but God does not for that matter create the world or relate to it by necessity. By His will God has chosen both to know and to accomplish what He knows. As such, the internal activity of God does not have any effects necessarily resulting from it if that should mean that they automatically originate from an immanent constraint of the divine being. To know the economy is an intransitive action on the side of God, but freely to will the actualization of what is known gives rise to external effects. This implies in addition that from eternity He has freely chosen not to refrain from creating the world. In the end, a consequence of this seems to be that God does *not* transcend all possible relationships with the world. I would, however, not draw this conclusion too hastily. It is one thing to relate internally to what is contemplated in the mind, it is another thing to relate externally to what is made according to the pattern contemplated. The first belongs to the divine sphere as such, the second to what is other than God. I think that according to Maximus, God, in His essential being, relates to nothing *other* at all. When He relates, the relation established is not an essential one, but it stems from the *activity* of the triadic being of God. Through the medium of the divine activity directed outside of

[143] *Cap. gnost.* 1.7, PG 90: 1085b.

the divine sphere, regulated by the divine will, the world originates. The sole bridge from the inner mystery of the divine being, perfect in itself, is the will to accomplish what is freely contemplated. This will is a manifestation of the *logoi*.

To make this clearer, we could formulate two propositions:

1. The internal activity of God is the necessary and sufficient condition for the existence of creatures.
2. The existence of creatures is incidental to the internal activity of God.

As far as I understand Plotinus, he would have conceded both propositions.[144] However, some remarks must be made regarding the second proposition. If it is correct to say—as I have above—that the One wills itself as good, i.e. as diffusive of good, then it could seem a bit strange to say that created being is *incidental* to the One's internal activity. On the other hand, to will oneself as diffusive of good does not necessarily mean to direct one's attention to something other than oneself. One could argue it is possible for the One both to will itself as diffusive of good and to be independent of the external result of this act of will. (An analogy could be that the will to go for a walk is independent of the footprints I make if I actually go for a walk.) Here the difference between Plotinus and Christianity comes to the fore, because the Christian God has chosen to direct His *attention* to the creation of something other than Himself. If we now look at the two propositions from the Christian point of view, we have to make some distinctions:

Ad (1): In the internal activity of God we must, at least formally, distinguish knowing as such from the act of will. We must also make a formal distinction between two aspects of the act of will. God's will to contemplate Himself as diffusive of good is the necessary but not the sufficient condition for the existence of creatures. What would make the contemplation a sufficient condition is the added formal aspect of will to create what is contemplated. With this, the difference

[144] On the subject of internal and external activity I am indebted to an article by E.K. Emilsson, 'Remarks on the Relation between the One and Intellect in Plotinus' (Emilsson 1999).

between Neoplatonism and Christian thought becomes still clearer: God has chosen (*a*) to know Himself as diffusive of good, and (*b*) to establish in actual fact through creative acts *ad extra* what is contained as a possibility in His knowledge. The Neoplatonists would not have conceded (*b*).

Ad (2): From a Christian point of view, we could at least distinguish between three kinds of activity in God: (*a*) the processions of the divine persons, (*b*) God's contemplative activity and (*c*) actions of God's will. One would have to say that the existence of creatures is incidental to the processions, but not to the actions of divine will. What, then, with God's contemplation? To the degree that God's contemplation is directed to the mystery of Christ as a possibility, the existence of creatures is incidental. To the degree that the will to create is added to God's contemplation of the mystery, the existence of creatures is not incidental. The realization of the divine economy is, then, dependent upon the *will* to accomplish it, and not on the contemplation as such. This once more brings into relief the difference between Christian and Neoplatonic philosophy because, from a Christian point of view, acts of will are necessary to bridge the gap between the divine knowledge and an actually existing world.

In my view, this is the position of Maximus. We have seen that God, according to Maximus, in His essence transcends relation. This implies that the eternal activity of God is *independent* of any relations to creatures. However, one aspect of this internal activity is God's knowledge of creatures. This knowledge or wisdom is eternal, and I believe it correct to say that it is God's contemplation of Himself as diffusive of good. According to Maximus, it is proper for the Good to diffuse or distribute itself in creative activity. This, actually, is something God wills.[145] The principles of this diffusion or distribution (διαστολή) are the *logoi* of beings, these *logoi* being identical with God's eternal knowledge of creatures.[146] In *Ambiguum* 7 St Maximus says that creatures are known in their *logoi*, but that this only represents a *possiblity* for their being created in actual fact.[147] Creatures are given existence, not by being known, but according to the wisdom of

[145] Cf. *De char.* 4.4, PG 90: 1048d, *Amb.* 7, PG 91: 1081c.
[146] Cf. *Amb.* 10, PG 91: 1177b–c.
[147] *Amb.* 7, PG 91: 1081a–b.

the Creator, at the appointed time. Here we have an implicit reference to the will of the Creator to accomplish what He knows eternally. As we shall see in the next chapter, the *logoi* themselves are not only God's knowledge of creatures, but function as divine acts of will at the moment for the actualization of the divine plan. As I said above, the distinction between God's knowledge and will is to be taken in the formal sense only.

Neoplatonist and Christian philosophers are often held to differ in their teaching on creation because Neoplatonism holds to a doctrine of emanation while Christianity does not. On this view, Christians teach that there is a basic gulf between created and uncreated being, while Neoplatonists think that there is some kind of continuity between the One and its creatures. I think it is wise to avoid charging Plotinus with a doctrine of emanation on the ground of his metaphors. Plotinus' philosophical teaching is more like a doctrine of creation than a doctrine of emanation.[148] The One is free and unrelated to anything else, Plotinus says.[149] The One acts as it wills, but acting thus, the *incidental* result is the establishment of the Intellect. The Intellect, on the one hand, is necessarily dependent on the One for its existence. The One itself, on the other hand, is not dependent on anything. If this interpretation is correct, there is obviously a basic gulf between the One and its effects. On the doctrine of creation, then, the primary difference between the Neoplatonist and the Christian positions has to do with differences in the understanding of God's intentions or will (whether or not He intends the creation of His effects).

According to Christian doctrine, God places a world 'outside of' Himself, as something *other*, separated from God by the basic gulf. The 'outside of' is not to be considered as something given prior to creation. There is nothing 'outside of' God in that sense. God creates the 'outside of' in the moment the world originates. The sense of this 'outside of' or otherness, is that *something* is instituted from *nothing*. In this way it is 'other' than God. The gulf that separates this 'something' from the divine nature is shown in the categorical limitation and essential determination of otherness, which opens up

[148] Cf. Emilsson (1999). [149] *Enn.* 6.8.8.

the possibility for movement in accordance with or contrary to the divine *logoi*. God, on the other hand, transcends all limitation or determination, and therefore all change. In the following chapters we will investigate more closely Maximus' understanding of the 'relation' between God and the world. This is a highly complex matter. We shall first discuss his famous doctrine of the *logoi*.

3

The Logos, the *logoi*, and Created Beings

So far I have abstained from translating *logos* (*logoi*), and I shall follow this practice throughout the rest of the book, for the question of how to translate this central term is a difficult one to answer satisfactorily. In English, as in other European languages, there is the long-established convention of translating the term λόγος in the prologue of the Gospel of John as 'Word' (German: Wort, French: Parole, Norwegian: Ord). However, I don't think it would be convenient to call the principles of beings in the Maximian philosophy 'words', even though this would take care of the linguistic (and semantic) connection between the Logos Himself and the *logoi* associated with Him. I don't think we are helped by the convention to translate the Stoic term λόγοι σπερματικοί as seminal reasons either. The term reason and reasons could of course be defined in an appropriate way, but 'reason' is used in other, non-technical connections and could cause confusion. Another, more obvious disadvantage is that reason lacks the immediate connection with the term Word in European languages. Since Word–words is problematic, and Word–reasons is inconvenient, I think it best to keep the Greek as technical terminology and speak of Logos–*logoi*.

The key to St Maximus' cosmology is his idea of *the mystery of Christ*. He develops this topic in broad lines in the *Quaestiones ad Thalassium* 60. Maximus comments on 1 Pet. 1:19–20, where it is said that Christ was known before the foundation of the world. Christ as foreknown is the same as the mystery of Christ. The Holy Trinity holds this mystery, in accordance with its essence (κατ' οὐσίαν).[1] It was foreknown, Maximus says, by the Father according

[1] *Ad Thal.* 60, CCSG 22: 79.105–8.

to His approval (εὐδοκία), to the Son according to His self-work (αὐτουργία), to the Spirit according to His cooperation (συνέργεια).[2] Christ was foreknown, not as God, but as man.[3] 'This is the mystery circumscribing all the ages revealing the super-infinite, great council of God, which in a manner beyond reckoning infinitely pre-exist the ages', Maximus says.[4] The divine purpose 'pre-exists' the ages, and Maximus' intention is clearly to stress that this is a mystery belonging to the sphere of God's eternal self-contemplation, beyond the 'before' and 'after' of the temporality of created being. The divine economy comprises both creation and salvation. In fact, creation and salvation are two stages in one and the same divine purpose: to make beings and to unite them with God. Beings are made with a view to deification. In this we perceive how the central motif of the Maximian system is the divine *philanthropia*, the great mystery that transcends all thought. *That* God eternally wanted to be the Creator and Saviour of the cosmos is a fact, but *why* is unanswerable. I think this gives additional emphasize to the point we touched upon earlier on creation (Ch. 2 § v): we may search to know the *how and whence* of creation, but not the *how and why* in a more profound sense.

The divine love is the essential mystery of Christianity. With this in view it is easy to understand the term Christocentric cosmology. The most profound meaning of the cosmos, its purpose and inner structure is situated in the triadic being of God, with Christ the Logos as centre of it all. It is in this light the doctrine of the *logoi* should be interpreted.

In the preceding chapter I claimed that Maximus' *logoi* of being are divine Ideas for all the things that have received their being from God. The theory of *logoi* is therefore a kind of Christian exemplarism worked out to suit the insight that God is the *free* Creator of everything. We have seen that, according to Maximus, the *logoi* are all contained in the divine wisdom from eternity; moreover, they are not only God's *thoughts*, but also His *acts of will*. In Chapter 2 § v we saw that he emphasized the role of the will in God's creative activity. In the *De divinis nominibus*, Dionysius the Areopagite called the *logoi* 'divine and good wills'.[5] In Maximus the identification of *logoi* and will is

[2] Ibid. 79.94–7.
[3] Ibid. 79.108 ff.
[4] Ibid. 75. 40 ff.
[5] *DN* 5.8, PG 3: 834c.

important because in this way a Christian alternative to Neoplatonic metaphysics comes to the fore.[6] He makes this identification for instance in the seventh *Ambiguum*. First he says that God knows all things by His *logoi* before their creation, and later on he qualifies this when he says twice that God knows all beings as His own acts of will (ἴδια θελήματα).[7] In *Ad Thal.* we find a reference to Dionysius when Maximus says that the *logoi* of beings were prepared in God before the ages and that they are 'customarily called good wills (ἀγαθὰ θελήματα) by the divines'.[8] The *logoi* therefore, are the Ideas through which the creative will of God manifests itself in the institution and ordering of visible and invisible creatures. We shall study this subject closely in the present chapter.

The doctrine of *logoi* is not only a doctrine of metaphysical principles in a cosmological sense, but is a multidimensional doctrine bearing on the whole divine economy of creation and redemption. Maximus' interpretation of the cosmos is a Christian one, and its basic point of orientation is the ultimate divine purpose (σκοπός) of the created totality. This end or purpose is the mystery of Christ:[9] 'For it was truly necessary that He who is by nature the Creator of the essence of beings should become in Himself, by grace, the Author of the deification of those whom He created; in order for the Giver of being to appear also as gracious Giver of eternal well-being.' The condition for deification is the Incarnation; yet the Incarnation results neither from a mechanical necessity within or without the deity, nor is it necessitated by original sin. The basic idea here is that *if* the divine goal—the deification of created beings, which have freely emanated from the divine Goodness—is to be fullfilled, *then* the Creator of nature must be the Author of deification as well.

The divine economy, according to Maximus, is expressed and fulfilled by a threefold presence of the Logos: in the cosmos, in Scripture, and in the historical person of Jesus Christ.[10] It is tempting to view

[6] Cf. the last part of Ch. 2 above.

[7] *Amb.* 7, PG 91: 1081a and 1085b.

[8] *Ad Thal.* 13, CCSG 7: 95.6–13.

[9] *Ad Thal.* 60, PG 90: 624d; CCSG 22: 79.117 ff. The text printed in CCSG reads εὖ εἶναι in line 119, so that the translation should be 'in order for the Giver of *well*-being' etc. According to the critical apparatus the εὖ is omitted in some manuscripts, and the logic of the sequence seems to demand its exclusion. Cf. also qu. 22, CCSG 7: 137.4–16.

[10] Cf. *Amb.* 33, PG 91: 1285c–1288a.

this threefold presence as a threefold incarnation.[11] Maximus speaks of the historical Incarnation of Jesus as 'His incarnational presence' (τῆς ἐνσάρκου αὐτοῦ παρουσίας), and when it comes to the Scriptures the Logos is said to 'embody' (σωματωθῆναι) Himself in them. Now, to speak of the presence of the Logos as 'incarnation' (ἐνσάρκωσις) or 'embodiment' (ἐνσωμάτωσις) in relation to the cosmos and to the Scriptures is a metaphorical usage of the terms. The cosmic embodiment is effected through the *logoi* of created beings, the scriptural embodiment through the *logoi* of Scripture. In the historical Incarnation *the person of the Logos* united Himself hypostatically with human nature.

This threefold scheme of 'incarnation' is a threefold and gradual revelation of the 'mystery of Christ'. If one is to understand the Christian metaphysics of St Maximus, the three moments should not be sundered, but must be held together. This, however, does not mean that it is impossible to focus on just one of them and make it the subject of closer study. What is important is not to isolate any one of the moments from the Maximian idea of a basic divine purpose of all creation.

In the present chapter I shall focus on the Logos as centre of all the *logoi* (§ I). I shall interpret the *logoi* as divine intentions for created beings, instituting an immanent order among them, viz. a sort of Porphyrian tree (§ II). This in turn will take us to the problematic of the principles of this ordering, that is to say the laws governing the relations of different beings to one another (§ III). The chapter will conclude with an investigation into the ontological constitution of created beings (§ IV). The discussion will lead us into the problematic of divine immanence and transcendence, but this important theme will nevertheless be postponed because its treatment presupposes the development of certain ideas in Maximus to which an entire chapter will be devoted (Chapter 4). These ideas pertain to his theory of the so-called divine activity or 'energy'. The full understanding of the divine *logoi* therefore, can only be harvested after we have discussed the relation between the *logoi* and the activity, that is to say in Chapters 4 and 5.

[11] Thunberg (1965), 82, talks about a 'three-fold incarnation'. He has, however, changed it to a 'three-fold embodiment' in the new edition of 1995, 77. Blowers (1991), 117 ff., has 'three incarnations'.

I. THE LOGOS AS CENTRE OF ALL THE *LOGOI*

A closer examination of the Maximian theory of divine *logoi* can take as its point of departure three texts in which St Maximus uses the image of the Logos as the centre of a circle and *logoi* as its radii. This is an expression of a basic metaphysical truth: the Logos is metaphysically the centre of the *logoi*, and He is present in each of them, as will be seen. The three texts are from *Cap. gnost.* 2.4, *Ambiguum* 7, and *Mystagogia* chapter 1. Maximus probably has taken this image from Dionysius the Areopagite, who, according to von Balthasar, most likely took it from Proclus.[12] Plotinus uses the image rather frequently, and therewith 'canonized' it for Neoplatonic thought.

Dionysius the Areopagite, the primary source of inspiration for this image in Maximus, uses it twice in his *De Divinis nominibus*, in chapter 2.5 (PG 3: 644a) and in chapter 5.6 (from 821a). In Dionysius, the images of the circle and of the sun illustrate several points: processions and *logoi* are united in God as 'centre' without internal confusion between the different processions or between the different *logoi*. He speaks, as a matter of fact of an unconfused union (ἀσύγχυτον ἕνωσιν) between the *logoi*.[13] Every procession (and every *logos*?) is a complete expression of the divine Goodness. By processions and *logoi* (as 'radii') God institutes creation by being somehow 'brought out' to it. A creature, for instance an animal, does not emanate from God. The creature is made from nothing and as made it participates in God as Life, Being, and Goodness. It does not participate in a *portion* of Life, Being, and Goodness, but in the whole of these gifts. Its capacity to receive, however, is limited by the *logos* or divine idea that defines its essence. The image of the circle points to a dialectic between unity and plurality that (however it is treated by Dionysius)[14] we will find again in Maximus as irreducible and God-willed.

In one way the circle-image in Maximus is easier to understand than in Dionysius, because what he wants to illustrate is quite

[12] Von Balthasar (1961), 593–4. [13] *DN* 5.7, PG 3: 821a/b.

[14] Cf. Spearritt (1968), 76, who contends that 'the priority of unity over multiplicity is emphasized, and God is proposed as the center of unity'.

unambiguous. The image clearly should tell us something about the relation between God or the Logos and the *logoi*.

(1) *Cap. gnost.* 2.4 (PG 90: 1125d–1128a): We shall begin with the text from *Cap. gnost.* because it is the most perspicuous of the three mentioned above. St Maximus says that 'just as straight lines which proceed from the centre are seen as entirely undivided (ἀδιαίρετος) in that position, so the one who has been made worthy to be in God will recognize with a certain simple and undivided knowledge all the pre-existing *logoi* of what has come into existence in Him (ἐν αὐτῷ).'[15] The radii are contemplated as undivided (ἀδιαίρετος) in the centre, and likewise the *logoi* are contemplated as undivided in God. The comparison of God and *logoi* with centre and radii shows us both (i) that the *principles* of created things are centred in God, and (ii) that God therefore is the centre of the *whole creation*. Later we will see that the circle-image harmonizes with two other fundamental structures of Maximus' metaphysics, viz. the dual 'movements' of procession and conversion (πρόοδος and ἐπιστροφή), and of expansion and contraction (διαστολή and συστολή).

The use of the term *undivided* is to be especially noted in the present text. It brings to mind the famous adverb (ἀδιαιρέτως) from the Chalcedonian definition of the Christological dogma. This is no coincidence, for sure. The 'undivided' should, however, be balanced with the 'unconfused' (ἀσυγχύτως) from the same definition, as is often done by Maximus. Here we meet a basic idea of his thought, the idea of unity in plurality. If the present text was the only one at our disposal, we might think that the *logoi*, which are differentiated in relation to one another when they 'proceed' from God, cease to be differentiated internally when considered *in* God. But this cannot be the case. If we consider the created world, plurality is not a temporary phase in the history of the cosmos, a phase which in the consummation shall be surmounted and transcended when

[15] Berthold in Maximus Confessor, *Selected Writings*, 148, and von Balthasar (1961), 593, seem to understand ϵν αυτω differently. Berthold understands it as a reflexive pronoun, αὑτῷ, which points to *the one who has been made worthy*, while von Balthasar views it as the pronoun αὐτῷ (αὐτός), which points to God. I find it most reasonable to believe that von Balthasar is correct. St Maximus talks about contemplation of the pre-existing *logoi* in God, and not in oneself.

everything arrives at an *undifferentiated unity*. Of course not every aspect of the cosmic manifold has the same degree of metaphysical value and permanence. To the degree that the cosmic plurality in 'this age' is ridden by sinful separations, it should be transcended; yet there exists an original metaphysical balance between unity and plurality that belongs to the created order as such, and this tension shall not be eliminated. If we move on to the level of principles, the created unity in plurality is secured by the relations between the *logoi* themselves. Because the foundation of cosmic plurality is a plurality of *logoi*, we have to understand that the unity of the *logoi* in God (as joined to the centre of a circle) is a differentiated unity; in other words, a unity in which the *logoi* are both *undivided* and *unconfused* in relation to one another. We shall return to this subject several times in what follows.

In the text we are considering, Maximus uses the expression 'pre-existing *logoi* of what has come into existence'. The *logoi* do not begin to exist at the moment of creation, but are 'pre-existent'. The expression 'pre-existing *logoi* of what has come into existence' can be compared with a related phrase in *Ambiguum* 7, which tells us that God possesses the *logoi* of what has come into existence before the ages were established.[16] God's possession of the *logoi* is equivalent to His *knowing* them. God knows all things before their coming to be, as we saw in the preceding chapter. When Maximus thus speaks about divine possession of the *logoi* before the ages, and of God knowing all things, he states that the *logoi* are *eternal* in the way that God Himself is. This is why in the *De charitate* he can use the expression God's 'eternally pre-existing knowledge of beings'.[17] As we saw in Chapter 2, God's knowing is related to His willing, so that the *logoi* may quite naturally be conceived as the condition of rational acts of will actualizing the plans or designs for created beings. As I pointed out earlier (Ch. 2 § v), there has to be at least a formal distinction between *logoi* as eternally known by God and *logoi* as acts of will in the moment of creation.

(2) *Ambiguum* 7 (PG 91: 1081c): Over this text we could place the following heading: the one Logos is many *logoi*, and the many are one. The one Logos is many *logoi* according to its creative

[16] *Amb.* 7, PG 91: 1080a.

[17] *De char.* 4.4, PG 90: 1048d: *Τὴν ἐξ ἀϊδίου ἐν αὐτῷ ὁ Δημιουργὸς τῶν ὄντων προϋπάρχουσαν γνῶσιν, ὅτε ἐβουλήθη, οὐσίωσε καὶ προεβάλετο.*

activity and its συνεκτικὴ πρόοδος, 'procession which keeps together'. The many *logoi* are one according to a 'converting transference' (ἐπιστρεπτικὴν ... ἀναφοράν) and 'hand-leading providence' (χειραγωγικὴν ... πρόνοιαν). In this way everything is led back to 'the all-powerful origin or to a centre which anticipates the beginnings of all the straight lines that come from it', as Maximus says.

According to Sherwood, the seventh *Ambiguum* is part of the Maximian attack on Origenism.[18] Maximus is commenting on a difficult text from one of the theological orations of St Gregory of Nazianzus, containing the words 'we, being a portion of God and flowed down from above'.[19] These words seem to have been appropriated by Origenists in defence of their doctrine of an original *henad* of intellectual beings around God, and their idea of a pre-cosmic fall. Now, in what way are the words of the Theologian to be correctly understood? In what way could intelligent beings be 'portions of God'? Maximus of course could not accept that a venerable Father was guilty of heresy. To him therefore, it was no option to understand these words in the sense that intelligences had existed as actual individual beings in two phases; in the first phase in an original unity around God, and in the second one as fallen beings in a cosmos created for their education to a virtuous life. He therefore considers that to be a portion of God simply means that beings have their cause in eternal *logoi* of different beings in God. In other words, these *logoi* are *divine Ideas* through which the essences of such beings are instituted by the creative act. This is the only possible way in which created beings could be termed 'parts of God'. The words 'slipped down from above', according to Maximus, have to be interpreted as referring to the historical fall of man, in which he lives not in accordance with his *logos* in God.[20]

Sherwood's suggestions are valuable. Maximus' project is anti-Origenistic, but it represents the development of a Christian metaphysical system that in fact becomes an alternative to Neoplatonic philosophy as well. This is affirmed by a close reading of the seventh *Ambiguum* itself, as will be seen below.

18 Sherwood (1955), 21 ff.

19 St Gregory of Nazianzus, *Oratio* 14.7, PG 35: 865c.

20 *Amb.* 7, PG 91: 1181c.

In the present text we find the double movement of 'procession that keeps together' and 'converting transference'. For simplicity I just use the terms *procession* and *conversion*. Neoplatonic philosophers use both these terms. Generally, one should avoid translating ἐπιστροφή as 'return' because according to Neoplatonic doctrine what is generated in the metaphysical process does not literally *go back* to the higher principles.[21] Rather what is generated *turns towards* the principle from which it came, and as a result of this 'turning towards' it receives some further ontological determination. This is in accord with Christian doctrine as well, because according to Christian beliefs beings do not literally return to be absorbed in the divine essence, even if they turn towards God and receive deification. These are the reasons why ἐπιστροφή should be understood as 'conversion'. However, if these restrictions are kept in mind, one could speak of 'return' as a metaphorical expression for conversion. In *Ambiguum* 10, Maximus employs the three prepositions ἐκ, ἐν, and εἰς to characterize the relation between created beings and God.[22] Everything has come to be *from* (ἐκ) God, is held together *in* (ἐν) Him, and 'everything will convert *to* Him' (εἰς αὐτὸν τὰ πάντα ἐπιστρέφεσθαι). This scheme reminds one of the well-known Neoplatonic triad of remaining–proceeding–converting, but also of Pauline formulas from the Letter to the Romans (11:36) and the Letter to the Colossians (1:16). In the *Ambiguum*-passage the prepositions ἐκ and εἰς indicate spatial movement, viz. movement *from* some centre and *towards* it. This clearly shows the metaphorical character of this way of speaking, since to come *from* God means to be created by Him, and to move *towards* God means that creatures exist for God's sake and therefore are made such that they naturally turn towards Him. To convert, however, is one of the possibilities connected with an earthly existence, and the actualization of such a potential is often described by Maximus as a kind of *movement*. Metaphorically speaking, this movement is a movement towards God, and in fact it culminates in the deification of the creature.

21 Cf. Lloyd (1990), 126 ff.

22 *Amb.* 10, PG 91: 1188b–c. We find a 'metaphysics of prepositions' in *Amb.* 7, PG 91: 1077c–1080a as well.

The theory of procession was used in later Platonism in an attempt to explain how unity gives rise to plurality. Plotinus frequently resorted to metaphorical explanations such as the image of the sun with its rays, and of snow with its coldness.[23] Behind these metaphors, however, we find the doctrine of double activity.[24] According to Plotinus, the One is the source of plurality or otherness just by being and remaining itself. The *genesis* of otherness does not in any way affect the unchangeable nature of the One. It remains forever in its own identity.[25]

We shall investigate the Neoplatonist doctrine of procession and conversion a bit closer. Proclus elaborated earlier doctrines of procession and conversion. In him we find the well-known idea that 'every effect remains in its cause, proceeds from it, and converts to it'.[26] The *remaining* is to be understood in the sense that the cause is the necessary and sufficient reason for the existence of the effect. The quality that produces and characterizes the effect is perfectly present in the cause. The *procession* accounts for the fact that cause and effect are different entities, while the *conversion* means that the effect is constituted as an entity by its turning towards the cause and by the reception of its quality from it. Every complete (*τέλειον*) being generates something, Proclus says, but remains itself undiminished.[27] This claim makes sense on the background of two Neoplatonic principles, viz. (i) that it belongs to the nature of what is complete, and therefore *good*, to distribute itself, and (ii) that the cause remains undiminished according to the logic of Plotinian double activity. The last principle secures the integrity and transcendence of the cause in relation to its effects. The cause produces what is like to itself (*τὰ ὅμοια πρὸς ἑαυτό*) and here we find another important principle, because this means that the participants receive *communion* with the nature of the cause.[28] It is necessary for the effect to participate in the cause because the effect receives its essence from the cause. This means that there is an intimate relationship of sympathy between effect and cause, since the effect is dependent upon the cause by nature, and desires (*ὀρέγεται*) union with it. The striving by the lower for the

[23] *Enn.* 1.7.1; 5.1.6. [24] As we saw in Ch. 2. [25] *Enn.* 5.1.6.

[26] *Elements of Theology*, prop. 35: *Πᾶν τὸ αἰτιατὸν καὶ μένει ἐν τῇ αὐτοῦ αἰτίᾳ καὶ πρόεισιν ἀπ' αὐτῆς καὶ ἐπιστρέφει πρὸς αὐτήν.*

[27] *Elements of Theology*, prop. 25–6. [28] *Elements of Theology*, prop. 28.

higher is supported by the goodness of the cause as something to be desired. The One, the sole origin of the whole cosmos, is identical with the Good itself.[29] In this way a universal cosmic striving for what is highest is grounded.

We have seen what is implied by the procession of effect from cause, and from this it should also be clear that *conversion* takes place on the basis of likeness.[30] What appears then is an image of cyclic activity, so that beginning and end (ἀρχή and τέλος) are linked together.[31] Everything proceeds like in a circle, from its causes and back to its causes again. Even though the ultimate source of everything is the One and Good, we may say that the object of every striving is the Intellect, because the whole cosmos has its essence from it. But the world-order, according to Proclus, is eternal. It eternally *proceeds* from the Intellect, is eternal in its own essence in which it *remains* according to its own order, and eternally *converts* to the Intellect.[32]

There are similarities as well as important dissimilarities between the cyclic 'movements' of procession and conversion in Proclus and in Maximus. Maximus' primary source for these structures is Dionysius the Areopagite. Dionysius, one believes, developed his Christianized version of Neoplatonic metaphysics from Proclus' doctrines. Maximus uses the theory of procession and conversion in a way similar to Proclus to explain how unity gives rise to plurality and how plurality is kept together in unitary fashion by the One. The idea of the One Logos as many and of the many as One points to a double 'movement' of the One.[33]

The double movement of procession and conversion for Maximus has, as we can understand, primarily to do with the institution of created beings *with* time and the final consummation of beings. The metaphysical structure of procession and conversion is the basic condition for the historical existence of beings, a point of primary concern in *Ambiguum* 7. Likewise in connection with conversion, the Providence which leads created beings 'by hand' (χειραγωγικήν ... πρόνοιαν) is mentioned, which seems to point to God's activity not only as final cause attracting beings to their related

[29] Ibid. 28, 25, 13. [30] Ibid. 32. [31] Ibid. 33.
[32] Ibid. 34. [33] Cf. *De char.* 4.4, PG 90: 1048d.

source, but to His grace by which He supports their turn towards Him. The conversion is not determined by some kind of natural necessity, even though beings *naturally strive* towards God. Intellectual beings are free to move in accordance with or discordant with their ontological principle (*logos*). The One Logos operates as an efficient and formal (paradigmatic) cause instituting a manifold cosmos through the many *logoi*, and likewise, by grace, He operates as a final cause, permitting the natural conversion of the manifold to unity.

We shall dwell for a moment on Maximus' Christian version of the three moments of remaining, procesion, and conversion successively:

(*a*) *Remaining*: According to Sherwood, Maximus revises Origenist doctrine when he substitutes the Origenist triad rest–movement–creation with creation–movement–rest.[34] The Origenist idea is that intellectual beings are originally gathered around God (this is *μονή*), but because of surfeit in their contemplation of God they move away from Him and thereby sin (this is *κίνησις*). As a result of this the world is created as a pedagogical institution for the redemption of intellectual creatures (this is *γένεσις*).[35] God, Maximus teaches, is the *archē* of *genesis*, and *genesis* is logically prior to movement. The condition of rest, or the immobile state (*στάσις*), is the final end of creaturely motion.

All of this sounds good, but there is one aspect of Maximus' teaching in the seventh *Ambiguum* that could be missed if one focuses solely on his refutation of Origenism. St Maximus obviously has a doctrine of rest as an *original* metaphysical condition as well. I have already indicated above how he interprets St Gregory's term that creatures are portions of God. In his own words:[36] 'Therefore, we are and are said to be a "portion of God" by the fact that the *logoi* of our being are established as pre-existing in God.' There is, then, an original remaining, but this remaining is not the Origenistic one of particular beings around God, rather it is on the level of metaphysical principles. Particular beings do not originally exist 'in' or 'around' God, but God *knows* in His *logoi* all the things that He will create. We find the same way of thinking in the Areopagite too: Dionysius

[34] Sherwood (1955), 92–3.

[35] Cf. Meyendorff (1975), 132–3, *Amb*. 7, PG 91: 1069a ff., cf. *Amb*. 15: 1217.

[36] *Amb*. 7, PG 91: 1081c: *Μοῖρα οὖν ἐσμεν καὶ λεγόμεθα Θεοῦ διὰ τὸ τοὺς τοῦ εἶναι ἡμῶν λόγους ἐν τῷ Θεῷ προϋφεστάναι.*

says that Being, and this means—I think—that the totality of *what is* is in God, but He is not in being. God is all things as the cause of all things.[37] The most plausible interpretation of these expressions is that creatures are in God because He conceives them in the divine *logoi*. God is all things because He thinks all things in these Ideas.

(*b*) *Procession*: The procession, in which the one Logos is many *logoi*, is the manifestation of God's eternally pre-existing *logoi* in the act of creation. Here a certain aspect of Maximus' terminology should be noted. We saw that the 'movement' from the one Logos to the many *logoi* the procession was qualified as *συνεκτική*. This is an interesting term. We are dealing with a causal relation (the act of creation), and the word *συνεκτική* was used in Stoic theories of causation. One of our sources for the Stoic concept of the *συνεκτικὸν αἴτιον*, is Galen (AD 129–?199). According to Galen, however, this cause is not the cause of coming to be (*γένεσις*), but of existence (*ὕπάρξις*).[38] The 'sustaining cause' ('the cause that keeps together') is a dynamic element that holds together the material elements of a body, for instance the breath which sustains natural bodies.[39] For St Gregory the Theologian there is a distinction between the efficient and the sustaining cause. God, Gregory says, is the creative and sustaining (*συνεκτική*) cause of all creatures.[40] We get the same picture from St Gregory of Nyssa when he says that the Logos is God's eternal Power that creates and sustains beings (*ἡ ἀΐδιος τοῦ Θεοῦ δύναμις, ἡ ποιητικὴ τῶν ὄντων . . . ἡ συνεκτικὴ τῶν γεγονότων*).[41] In Proclus we find that unity is what conserves and *holds* each essence *together*. The Good, as the principle of unity, is sustaining of beings.[42]

It seems to me that for Maximus the concepts of efficient and sustaining causality are brought closely together, more closely perhaps than in the two Cappadocians. In *Ambiguum* 7 the sustaining procession obviously has to do with God acting as efficient cause in the creative act; but an element of this causal act is that God, the Giver of being, is at the same time God who preserves beings in their being. This indicates that the creative power, in accordance with the *logoi*, keeps not only individual creatures but also the created cosmos as a whole together as an ordered structure. Every region of the created

[37] *DN* 5.8, PG 3: 824a–b.
[38] Long and Sedley (1992), 1: 335–6/2: 336, H Galen.
[39] Ibid. 1: 334–5/2: 334–5, F Galen.
[40] *Oratio* 28.6, PG 36: 32c.
[41] *Oratio catechetica* 5, PG 45: 21b.
[42] *Elements of Theology*, prop. 13

world and every part within it is kept together by God who has established certain ontological bonds or laws which connect every being to every other in a hierarchic and harmonious arrangement. For Maximus the double movement of expansion and contraction is an important structure in this arrangement. We shall turn to this double movement shortly.

(*c*) *Conversion*: In the difficult words of St Gregory the Theologian it was said that we have 'flowed down from above'. According to Maximus, this is to be taken in the sense that we have not allowed ourselves to be moved by the *logos* according to which we were created.[43] The movement in discord with the *logos* causes sinful separations and enmity in the cosmos. The conversion then, is to be understood as the opposite movement, viz. as a movement *in accordance with* the *logos* of being.[44] It is said that in this last movement the many *logoi* are one Logos. This means that sinful separations are annihilated and beings are joined together in community according to the possibilities established by the *logoi*. Individuals are joined in their species, and species in their genera, and ultimately all things are joined together in community with God. The exact relation between the *logoi* and these individuals, species, and genera will be further determined below in this section and in later paragraphs of the present chapter.

Once again a special aspect of Maximus' terminology should be noted. The conversion is described as an ἐπιστρεπτικὴ ἀναφορά. The word ἀναφορά could just mean a 'leading up' or a 'transference'. But when used by this author it is tempting to detect a deeper sense, viz. a reference to the Eucharistic prayer (the *anaphora*). This may seem to be a somewhat speculative interpretation, yet at the very least it does not betray the basic convictions of Maximus. The universal cosmic purpose is a kind of Eucharistic communion in which every being is brought into the divine sphere to be deified. The whole cosmos, as we shall see, is to be actualized according to the good potential given to every being by God, in the universal 'Kosmische Liturgie'.[45]

The double movement of expansion and contraction, which is described in the tenth *Ambiguum*, should also be connected with the movements of procession and conversion.[46] I do not believe that the two pairs of movements are identical. They could, I suppose, be

[43] *Amb.* 7, PG 91: 1081c. [44] Cf. *Amb.* 7, PG 91: 1080c.
[45] Cf. *Myst.* 24, PG. 91: 709c. [46] Cf. *Amb.* 10, PG 91: 1177b–c.

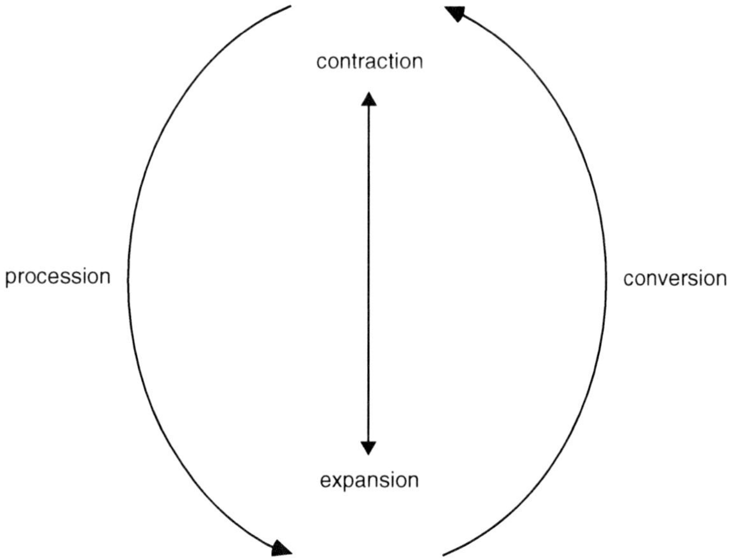

Figure 1. An illustration of the two double-movements of procession and conversion, expansion and contraction

understood that way, but on the basis of the texts in the *Ambiguum* 7 and 10, it seems more reasonable to think that procession–conversion is the movement establishing the historical limits of the world, while the movement of expansion–contraction is a kind of ontological arrangement or 'movement'. If this is correct, then the *one* double movement of procession and conversion in Neoplatonism, is modified by Maximus into *two* double movements, a modification which could be motivated by the Christian idea of a *historical* beginning and end of the cosmos. I think that the metaphysical process of expansion–contraction should be seen within the cosmological limits set by procession–conversion (see Fig. 1). It is obvious that procession points to the origin of creatures in God, i.e. to the *archē* of every existing thing in the creative act. Conversion, on the other hand, points to the final consummation of beings, i.e. to their final end in God. These limits circumscribe the double movement of ontological constitution, i.e. of expansion–contraction.

In the process of expansion, God, by the *logoi* of specific and generic being, distributes essences from the highest to the lowest kind of beings. This distribution culminates in the concrete plurality of created particulars. In the contractive movement, on the other hand, created beings are brought together in community within species and genera and in the end are unified in the highest *logos* of essence (οὐσία). The details of this 'Porphyrian tree' will be considered later in (§ 11). Expansion and contraction are *simultaneous* 'movements', pointing to the basic ontological truth of how everything created is by expansion defined by the formal circumscription of essence in a *differentiated* hierarchic system, and how creatures in contraction are gathered together from below and systematically established in *community* with each other in the same hierarchic system. This vertical relation downwards and upwards is, so it seems, an expression of formal and final causality by which creatures are constituted in their being.

It is very important to note that diversity as such is not something negative, something to be abolished in the conversion. The diversity of creation is established by the One Logos in accordance with His many *logoi*, and this selfsame diversity is instituted as a God-willed and irreducible unity-in-plurality. Between the processes of creative procession and redemptive conversion there opens up the *distension* (διάστασις) of historical time and the drama of human existence in its positive or negative relation to its basic *logos*.[47]

Let us sum up what is to be gathered from the circle-image in the seventh *Ambiguum*. The image tells us about a created order that has its beginning and end in the Logos as an almighty and gathering centre of all things through *logoi* as creative principles. A created arrangement is revealed, which because of the Logos/*logoi*-structure is manifested as a real and irreducible unity-in-plurality. The One Logos in the creative act should not be considered an empty name for a sum of *logoi*, nor is the Logos divided and distributed in portions. It seems rather that the One Logos, the second person in the Holy Trinity, holds the *logoi* together as His own plans or acts of will, and acts

[47] Διάστασις is a central concept in the thought of St Gregory of Nyssa, characterizing the cosmic condition under which created beings exist. Cf. von Balthasar (1995), 28 ff. Cf. the typical Gregorian text in St Maximus *Cap. Gnost.* 1.5, PG 90: 1085a. On human existence in accord or discord with its basic *logos*, cf. *Amb.* 7, PG 91: 1081c.

according to them in His creative and redemptive scheme *ad extra*. The text in *Ambiguum* 7 adds a further qualification to this image: 'Always and in all God's Logos and God wills to effect the mystery of His own embodiment.'[48] Considered in connection with the circle-image this means that through the *logoi* the Logos is 'embodied' in creation, or, to use a further metaphor, *incarnated*. This is to say that in creation is effected an 'incarnation' of the Son of God *before* His historical Incarnation. If there is an incarnation in the cosmos, then the relation between Logos and *logoi* has this additional character that each *logos* is able to represent *the Logos Himself* to the actual being in question. We will return to this subject below, by the end of § IV.

(3) *Myst.* chapter 1 (PG 91: 668a–b): In this text, which partly repeats what we have already seen in the foregoing two, the image of the circle is developed into an illustration of how Christ as God encloses in Himself (*ἑαυτῷ περικλείων*) all things by His one, simple, and infinitely wise power. This enclosure is emphasized when it is said that God, like a circumference, circumscribes the extensions of the principles of beings (*αἵ ἀρχαί τῶν ὄντων*) and brings them back to Himself as straight lines that are fastened on to the centre. Here we gain a vivid expression of divine Providence. The *principles of beings*, I suppose, are reasonably interpreted in accordance with the two texts we have just commented on, as *logoi of beings*. These principles are the means by which the Logos of God extends to the end of the world of creatures, like radii to the periphery of the circle, so that the creatures of God shall not be strangers and enemies to one another by having no centre or no place to show each other friendship, peace, and identity (*ταυτόν*), as the text has it. The principles are the divine means to establish metaphysical relations between created beings and Himself. Every cosmic movement, be it in accord or in discord with the basic 'logic' laid down by God, is somehow within the limits of His concern (i.e. Providence). The framework of nature itself guarantees a positive possibility even for the unruly creature not to risk a fall into non-being. Again we catch the vision of a created order firmly established in God's grip, because of the fundamental metaphysical framework and the ontological principles at work in it.

[48] *Amb.* 7, PG 91: 1084c–d.

God is both circumference and centre, all of which is accomplished through the *logoi* of creatures. The *logoi* therefore are the laws of the cosmos, not laws of nature in the modern sense of the word, but laws to be understood as divine efficient–formal–final causality binding together created beings horizontally and connecting them vertically with the Creator without transcending the basic differences between them or violating the border that has to remain between the created and uncreated orders. In this way the *logoi* are *logoi* of providence as well, and as such they become *logoi* of judgement when creatures move in relation to them—more on this later.

II. THE *LOGOI* AS PRINCIPLES OF A PORPHYRIAN TREE

We have seen that through *logoi* the Logos institutes an ordered cosmos. The next question we have to consider concerns the cosmic organization that this institution accomplishes. What *kind* of ordered system is brought into being? Central to it is the idea of a Porphyrian tree, in which beings can be grouped together in a hierarchical system of species and genera under the basic category of essence (οὐσία). In connection with the Christian metaphysics of St Maximus, for the most part I shall substitute the term 'Porphyrian tree' with 'taxonomic system'. 'Taxonomy' should in this context be understood as a classificatory system, not as a hierarchic system of values.[49]

The roots of the Porphyrian tree are, as we have seen above (Ch. 2), very old. But the full-fledged structure is worked out by Porphyry himself in the *Isagoge*. He says that in each category we find the highest genera, the lowest species, and some classes between these.[50] Now, this is a description of a way to classify every existing thing according to the characteristics of its essential being. Figure 2 is an example that illustrates the idea; I have excluded individuals from my

[49] In Lovejoy (1978), 64–5, a scale of beings is understood as a scale of perfections and, by implication, a scale of values as well. Such a scale of values is not intended here.

[50] Porphyrii *Isagoge*, CAG 4: 1, 4.14–20. Trans. Warren in Porphyry the Phoenician: *Isagoge*, 35.

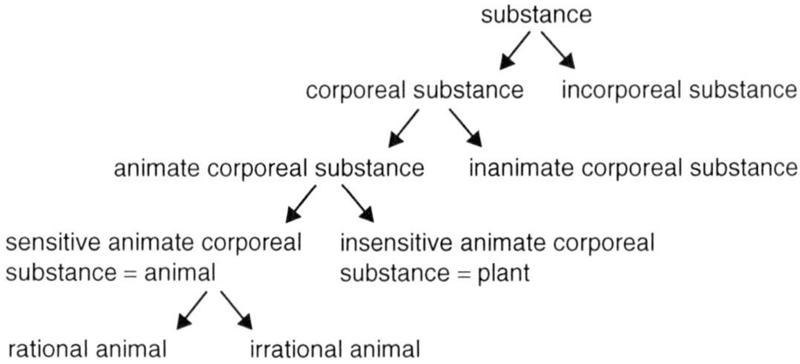

Figure 2. An example of the Porphyrian tree

diagram; however, under rational animal comes man and under man comes Socrates, Plato, i.e. particular men.

We have seen that in later Neoplatonism, for instance in Dexippus and Ammonius, this classificatory system was divided into two, viz. in a transcendent system and an immanent one (cf. Ch. 2 § III). The transcendent system is the Ideas in the Intellect, while the immanent one is the system of the sensible world that is instituted with the transcendent system as intelligible paradigm.

What then of Maximus? In *Ambiguum* 41 Maximus comments on the following sentence from St Gregory the Theologian: 'natures are instituted afresh, and God becomes man.'[51] Maximus points out that in the Incarnation God has accomplished the cosmic task which man abandoned, and restored human nature to its original function. This *cosmic* role involves the unification of every created division of being, and for this reason Maximus depicts two such systems of division. The first, at the beginning of *Ambiguum* 41, is the famous fivefold division which Thunberg interprets in detail in chapter six of his *Microcosm and Mediator*. The second one is described near the end of the text, giving the whole of *Ambiguum* 41 a kind of symmetrical appearance: first the fivefold division is outlined, then the human role is described, the fall of human beings, the Incarnation and what is accomplished through it, and finally the whole cosmic system is outlined once more, but this description is different from the first one.

[51] *Sermon* 39.13, in PG 36: 348d.

The fivefold division by Maximus is of being into uncreated and created nature, of created nature into intelligible and sensible nature, of sensible nature into heaven and earth, earth into paradise and *oikoumenē*, and *oikoumenē* into male and female. Maximus' second division near the end of *Ambiguum* 41 however, is not only a successive division of creation into vast *fields of being*, but moreover delineates the contours of a detailed classificatory system based on a nuanced application of principles. I shall treat of the refinements of the system in § III.3 below. What is explicitly brought forward by Maximus is the role of the Logos and His *logoi* in the institution of the system of creatures. In this way he is able to show in which way the Logos Himself is both the Origin of cosmic differentiation and the One in whom everything is held together. Thus he demonstrates how the One Logos was able to recapitulate everything in Himself.

In the second division of being, as presented in the text, we find a system of classification that at the same time is an ontological system of the real world.[52] At the bottom of the system the accidents (τὰ συμβεβηκότα) are brought together with one another and given unity in the subject (τὸ ὑποκείμενον). It is not obvious what this means, and it calls for further comment. I believe that the 'subject' in question refers to particular beings of which 'accidents' are attributes or predicates. The subject is equivalent with the Aristotelian primary substance of the *Categories*, and the accidents are the predicates of this substance.[53] To call *all* predicates accidents is not ordinary Aristotelian usage, however. The accident is usually understood as one of several predicables.[54] Aristotle mentions four kinds of them: definition, property, genus, and accident. In the *Topics* he ranks difference together with genus.[55] In Porphyry's *Isagoge* there are five predicables: genus, difference, species, property, and accident.[56] What is notable here, is that difference is brought forward in its own right and that species has taken the place of definition, which is quite in

[52] *Amb.* 41, PG 91: 1312c–d. E. Perl (1991), 166 with n. 46, makes the observation that this text has received little attention from modern commentators. It is neglected by A. Riou (1973). Thunberg (1965, 66–7; 2nd edn. 1995, 63), von Balthasar (1961), 156–7, and Gersh (1978), 141 n. 80, mention it 'with regard to Maximus' theory of differentiation and union, but only briefly and without drawing attention to its full significance'.

[53] *Cat.* 1a20 ff.

[54] Cf. *Topics* book 1, chs. 4–5.

[55] *Topics* book 1. 4.101^{b}17–19.

[56] Porphyry, *Isagoge* CAG 1.1.3–5.

order. In Aristotelian thought, accident is the predicable that in a special way denotes the individual as such. Aristotle however, in at least one place in his *Metaphysics*, feels free to include both accident proper and property (τὸ ἴδιον) under the heading of 'accident'.[57] What type of predicates Maximus intends to include under the heading of accidents (συμβεβηκότα) is not clear; perhaps he includes both accidents and properties to the exclusion of essential predicates, as does Aristotle himself. If this is so, then the kind of being at the lowest level of Maximus' system could be understood as *a qualified individual*, which, as we soon shall see, is identical on the higher level with other qualified individuals in a common species.

The elementary logic behind Maximus' statements could well have been learnt in his youth or during the time when he was head of the imperial chancellery under the emperor Heraclius. One of the logical compendia that he probably made use of contains sections on the five predicables of Porphyry.[58] On the other hand, even if we suppose that he in fact did study these philosophical texts, it is nevertheless probable that Gregory of Nyssa's *De hominis opificio* is the immediate source of inspiration behind Maximus' ideas about subject and accident. There are at least three passages in Gregory's works in which he develops his concept of matter.[59] The only one in which he mentions a subject (ὑποκείμενον) is from *De hominis opificio*. In this subject all kinds of attributes come together. Examples of such attributes are colour, weight, quantity, qualities of touch, figure, resistance, extension, etc. I believe attributes like these could be Maximus' 'accidents'. In connection with this, however, three points should be made, the first one concerns the nature of the subject, the second is terminological, the third one concerns the relation between the basic level of *subject and accidents* with the next level pointed out by Maximus, that of *individuals*. (i) I doubt that Gregory thinks of the subject as a 'something' with a separate existence. The subject, if we take account of all three texts, is probably to be understood as identical with the bundle of attributes. (ii) The terminological

[57] *Metaph*. Δ, ch. 30.

[58] Roueché (1974), 74–5 and Roueché (1980). Cf. § III of the Introduction.

[59] *In hexaemeron*, PG 44: 69b–c; *De anima et resurrectione*, PG 46: 124b–d; *De hominis opificio*, PG 44: 212d–213c. Cf. the discussion in Sorabji (1988), 52 ff. Chapter 4 in Sorabji contains an interesting discussion on 'Bodies as bundles of properties'.

point occurs because Gregory in none of these texts uses the term accidents, rather his terms are ποιότητες and ἰδιώματα, respectively 'qualities' and 'properties'. On the other hand it seems quite in line with St Gregory when in *De charitate* Maximus states that God created the essences (οὐσίαι) endowed with qualities.[60] In *Quaestiones et dubia* 104 he actually treats the terms accidents and qualities as synonymous.[61]

(iii) The third point is, perhaps, more difficult to sort out. Gregory speaks of the constitution of the basic subject of the world, i.e. about the way the substance or material stuff of beings is built up.[62] Is this what Maximus is concerned with? I think not. As we will soon see, he is describing the division of beings *in accordance with the divine logoi*. The *logoi* are principles that are institutive of the essences of creatures. Consequently, the subject with accidents in *Ambiguum* 41 is a level of essential being below the level of the species, not matter as such.[63] But if Maximus is not speaking of matter as such, he might be distinguishing between the subject and the individuals (τὰ ἄτομα).[64] But are they really to be understood as *two* different essential levels? Could it not be that the subject and the individual denote one and the same thing, but this thing is viewed differently? Viewed as *subject* the particular essence is understood as a concrete instantiation of being in the world, i.e. at the level at which it contracts all kinds of external predications. Viewed as *individual* this same instantiation of the essence is understood in relation to the species. I suppose this is a probable interpretation. However, it is at least highly interesting that Maximus actually teaches the existence of essences of *particulars* with *logoi* of their own.

In accordance with my interpretation, on the next level, the subjects, understood as qualified individual beings, are considered to be identical with each other in the species (τὰ ἄτομα δὲ κατὰ τὸ εἶδος ... ταὐτὸν ἀλλήλοις), and, further on, the species receive

[60] *De char.* 4.6, PG 90: 1049a. [61] *Qu. Dub.* 104, CCSG 10: 78.5 ff.

[62] St Gregory defines the hypostasis in this way as well, cf. Pseudo-Basil, *Letter 38*, 218 (Loeb). See my treatment of hypostasis § IV below.

[63] This may be a weak argument. I admit that I could be wrong. I hold the possibility open that the crux of the subject could be the concrete stuff of the world, its matter. If that is so, it is on this material basis that essentially determined beings are built up.

[64] *Amb.* 41, PG 91: 1312d.

identity with each other according to the genus (*τὰ δὲ εἴδη κατὰ τὸ γένος . . . τὴν πρὸς ἄλληλα ταὐτότητα δέχεται*). Even though Maximus does not complete his theoretical account of this system by explicitly bringing in intermediate genera/species, he of course does not exclude them in principle. This contention is confirmed by a text in the tenth *Ambiguum*, where he, when describing the system, uses the terms the most generic genus (*τὸ γενικώτατον γένος*) of the highest genus, speaks of intermediate genera, and even of the most specific species (*τὰ εἰδικώτατα εἴδη*)—which, according to Porphyry, is the lowest level of universal being, the lowest species.[65] There are several levels of intermediate genera, of more and less *γενικώτερα γένη*. This text will be examined more closely in the next section.

Now, in what way is the system of individuals, species, and genera related to the *logoi* of the Logos? The context of the system described above (*Ambiguum* 41) seems to indicate an answer to this question.[66] Maximus speaks of *a generic logos of nature* which forces together what is differentiated to 'the one and the same'. He further says, 'everything generic, *according to its own logos* (*κατὰ τὸν οἰκεῖον λόγον*), is wholly present . . . to those subordinate wholes'. Then he says that the *logoi* of that which is universal and generic 'contains the *logoi* of everything that is divided and particular, as they say'. I interpret these phrases to mean that the *logoi* are *the divine principles by which individuals, species, and genera are instituted in a created hierarchic system of essences*. We have then to distinguish between the essence of the individual (for instance of Peter's manhood), the specific essence (man) and the generic essence (mortal animal) on the one hand, and *the logoi of* individual essence, specific and generic essence on the other. The taxonomic system introduced in this text is first and foremost *the immanent order of created beings*.

At this stage we should address a question that naturally turns up, and at the same time sum up in a systematic way what we have learnt about the *logoi* up till now.

Chapter 2 § III on exemplarism in the Platonic tradition ended with Dexippus and Ammonius, both of whom taught that universal species and genera exist in the Intellect as paradigms for sensible

[65] *Amb*. 10, PG 91: 1177c. Cf. Porphyry, *Isagoge*, CAG 4: 1.4.15 ff.
[66] *Amb*. 41, PG 91: 1312b–1313b.

things. It would be quite natural to ask if Maximus held a similar view, and it comes as a surprise to find him saying that universals are created.[67] He also says that since universals consist of particulars, if the particulars perish, then the universals perish with them.[68] The *logoi* are not creatures and are definitely not among the things that perish with the destruction of individual creatures. As far as I know, he never talks of the *logoi* as universals, and against the background of Neoplatonic thought all this seems strange. How should it be explained? St Maximus tells us that the knowledge or wisdom (*γνῶσις* and *σοφία*) which God has of created beings, is in Him from all eternity.[69] I think we can say that this eternal wisdom is identical with the sum total of all the *logoi*. God knows in these *logoi* what He will make. He has defined beings and He wills their creation. Here we should remember a distinction I felt it necessary to make in Chapter 2 § v on Maximus' doctrine of creation, (i) between the will of God to be Himself and contemplate the *logoi* as His eternal wisdom, and (ii) the *logoi* as acts of will at the moment of creation. What He has defined (eternally) and what He wills (at the moment of creation) is conceived in the *logoi* as a system of essence (*οὐσία*) with internal differentiations and identifications on the different levels. God therefore *knows* the taxonomy through the *logoi*, but the *logoi* themselves are the principles of the system. They are not taxonomically arranged universal genera and species in themselves. God's knowledge is simple and unitary. In knowing Himself, He knows all possibilities.[70] This

[67] *Amb*. 7, PG 91: 1080a. [68] *Amb*. 10, PG 91: 1189c–d.

[69] *De char*. 4.3–4, PG 90: 1084c–d.

[70] Cf. *Amb*. 7, PG 91: 1081a and c; *Myst*. ch. 1, PG 91: 668a. Törönen (2007), 130 n. 39, seems to criticize my interpretation of the *logoi*, the Porhpyrian tree, and the divine Ideas. It is not easy, however, to see exactly what is criticized. It should be quite clear that I distinguish between the *logoi* and the taxonomy, that *logoi* are divine principles, while the taxonomy (including species and genera) is the immanent order of the cosmos. I do not identify *logoi* with Platonic Forms the way such Forms are conceived in the history of ideas. Törönen further remarks that the identification of Platonic Forms with the species of the Porphyrian tree is unjustified. Yet later Neoplatonists like Dexippus and Ammonius do not seem to share his view—perhaps one's sympathy or antipathy with the identifications and differentiations in this connection depends upon what one understands a Platonic Form to be. However this may be, Maximus would naturally conceive that what God contemplated in the eternal plan of His wisdom, as the sum of *logoi*, somehow would be divine conceptions (i.e. definitions) of beings, conceptions which in the next step become acts of will 'when' He creates the cosmos.

eternal wisdom is manifested in a plurality of *logoi* at the moment of creation. The divine being itself is not to be identified as a reservoir of Ideas or a container of static Forms, and this is why I have called Maximus' doctrine of *logoi* a *kind of* exemplarism. What strikes me as important is to understand the *logoi* first and foremost as the *principles* of an immanent taxonomic order of genera and species. However, there is one more challenge to be addressed in connection with this problematic.

Even if the *logoi* are not Forms that may be arranged in a static taxonomy in the divine intellect, the *logoi* are God's plans or definitions or predeterminations (*προορισμοί*) of what He will create.[71] When I commented on the text from *Cap. Gnost.* 2.4 above in § 1, containing the circle-image, I stressed two things, both of them important: (i) the unity of the *logoi* at the centre, and (ii) the differentation of this unity as such, as an expression of the essential divine wish to make and preserve plurality. On the one hand, we can see from the texts that Maximus wants to stress the principle of divine simplicity. This is witnessed by his use of the circle-image and is, for instance, expressed in *Mystagogia*, when he talks of 'the unique, simple and infinitely wise power of His goodness'.[72] Somehow, both unity and differentation is present at the highest level of reality. How? The Plotinian Intellect contemplates the Forms as its thoughts, and there is unity because what contemplates (subject) and what is contemplated (object) are the same. Of course, this is the case with God and His divine wisdom expressed in the *logoi* as well. I suppose we could say that God knows as Logos, in a simple way, His own perfection as paradigm of a possible multitude of created beings. This could also be seen from another angle. In *Ambiguum* 41, Maximus says the *logoi* of individuals are contained in the *logoi* of species, the *logoi* of species by the *logoi* of genera, and the highest *logoi* are contained in Wisdom, i.e. in Christ.[73] Here there are two things to be noted. On the one hand, the *logoi* are arranged, the one in the other hierarchically. This shows that in one act of knowledge, knowing His own Wisdom, i.e. knowing Himself, He knows all the principles (*logoi*) for all the things He wants to make. On the other hand, the text *could* be interpreted to indicate a

[71] *Amb.* 7, PG 91: 1085a. [72] *Myst.*, ch. 1, PG 91: 668a.
[73] *Amb.* 41, PG 91: 1313a–b.

taxonomy of *logoi*: the lower contained orderly in the higher. However that may be, the important thing is to understand the *logoi* as the principles of created arrangements (universal genera and species) and individuals.

In Scholastic philosophy one finds the theory that the divine intellect contains a great many Ideas from which God has selected a number and combination in order to create this our world. According to Thomas Aquinas, God knows infinite things.[74] A divine Idea can be seen either as an *exemplar*, that is to say a 'principle of the making of things', or as a *ratio*, which is a 'principle of knowledge'.[75] Thomas says: 'As an exemplar, therefore, it has respect to everything made by God in any period of time; whereas as a principle of knowledge it has respect to all things known by God, even though they never come to be in time.' All the divine Ideas are principles of knowledge, but God chooses some of them as principles of making. The distinction between the Idea as principle of knowledge and exemplar is similar to the Maximian distinction between what God knows eternally and how this knowledge becomes acts of will in creation. If we return to Thomas this enables him to argue that God, in His infinite intellect, knows an infinity of Ideas from which He selects a finite number and combination so as to create one among a number of possible worlds. As it is, however, Thomas denies the existence of possible worlds on Aristotelian principles.[76] On 7 March 1277, the bishop of Paris, Etienne Tempier, issued the famous decree that condemned 219 propositions of a philosophical nature.[77] One of the erroneous propositions was 'That the first Cause cannot make several worlds'.[78] Had he still been alive, Thomas Aquinas could have defended himself along the lines hinted at above, even though he would have had to revise some of his Aristotelian tenets.

Maximus seems to think otherwise. We should remember what was said in Chapter 2 above: God not only chose to know the mystery of His embodiment in created being, He even decided to accomplish it. According to Lovejoy, in his famous *The Great Chain of Being*, such a

74 Cf. *ST* I, q. 14, art. 11.

75 *ST* I, q. 15, art. 3 (*Summa Theologica*, trans. Fathers of the English Dominican Province 1911; repr. 1981).

76 Cf. Duhem (1985), 447 ff.

77 Cf. Gilson (1980), 405 ff.

78 Cf. Duhem (1985), 449 ff.

doctrine could be seen as pointing to one of the tensions in Christian thought, because 'the whole tendency of the Neoplatonic dialectic is adverse to that conception of arbitrary volition and capriciously limited selection from among the possibilities of being, which was to play a great part in the history of Christian theology'.[79] In his study Lovejoy defined 'the principle of plenitude', according to which 'no genuine potentiality of being can remain unfulfilled' and said that 'the extent and abundance of the creation must be as great as the possibility of existence and commensurate with the productive capacity of a 'perfect' and inexhaustible Source'.[80] This principle, Lovejoy holds, is connected with the idea of the natural necessity of creation.[81] We saw in Chapter 2 that according to Maximus, God is not subject to necessity. The free divine will plays the decisive role in the creation of the world. It is, however, difficult to agree with Lovejoy that this discloses an idea of the divine will as something *arbitrary* and *capricious*. Maximus does not seem to be interested in the philosophical problems related to questions about the infinite number of divine Ideas and the existence of possible worlds. God, without any external or internal constraint, has somehow limited Himself in accordance with a certain purpose. His Goodness has freely taken the shape of love for man (φιλανθρωπία) with the limited scope of accomplishing the economy He conceived before the ages. God, in His eternal being, has by sheer goodness chosen to be centred upon the Idea of human nature as the natural bond (σύνδεσμος) which shall freely connect levels of being on to Himself.[82] The world is not just one among an infinite number of possible worlds, but is the cosmos which in its original, natural purity is an expression of God's choice to be the all-powerful lover of men and the one who bestows the gift of deification. The doctrine of Logos–*logoi* is an expression of 'the mystery of Christ'. Christ Himself is not only the transcendent divine Logos, but is the incarnated one. This divine condescension is a natural act of God who from all eternity remains constant in His intentions. Even though there might be problems connected with the idea of a good and free creator, I do not think that there is a tension in Maximus' thought of the kind Lovejoy envisages. My interpretation

[79] Lovejoy (1978), 63. [80] Ibid. 52. [81] Ibid. 54.
[82] Cf. *Ad Thal.* 22, CCSG 7 and *Amb.* 41, PG 91: 1305a ff.

of Maximus' thought is confirmed by what he says in the *Ambiguum* 42:[83] 'For none at all of the beings exists, the *logos* of which does not at all events pre-exist with God. Of the beings which *logoi* of essence are pre-existing with God, of these manifestly, there assuredly are coming to be, according to divine council.' The divine plan is a definite one, limited to the accomplishment of one central purpose: the mystery of Christ. I believe Thomas Aquinas himself would have sympathized with this doctrine, even though his doctrine of divine Ideas opens up a way of philosophizing that would not have interested Maximus much.

In the Neoplatonic Intellect, the Ideas may be seen as an established taxonomic system of universals. The influence of St Augustine and of the Latin translation of Porphyry's *Isagoge*, connected with the discovery of the Aristotelian *Corpus*, made the idea of such a taxonomic system of universals appear quite natural to Scholastic philosophers. To Maximus, on the other hand, the *logoi* do not really seem to be universals *in themselves*, but are rather *principles* of immanent universal arrangements. The divine Logos manifests from Himself a *logos* of being (*οὐσία*) as universal category, *logoi* of genera and species, *logoi* of individuals; all of which *logoi* are *principles* of universal and particular being. I quite disagree with Eric Perl, who says that the Word of God is the highest universal, and that it as such contains all the other *logoi*.[84] It seems to me that Perl is of the opinion that the *logoi* are themselves organized as a taxonomic system of universals, even though he is not clear on the matter. He says, 'The logoi are the principles whereby universal Being is appropriately distributed to each creature so that it may be.'[85] 'Being' here is not essence (*οὐσία*) as the highest category of immanent essence, but rather a divine perfection or attribute. In this context we should keep in mind the distinction mentioned earlier between *logoi* and activities (*ἐνέργειαι*), which will be more fully developed in my next chapter (especially in Ch. 4 § IV). I agree with Perl that the *logoi* are the principles whereby Being is distributed. But when he next says the 'perfections are arranged in a hierarchical order according to degrees of universality, while the *logoi*

[83] *Amb.* 42, PG 91: 1329b: *Οὐδὲν γὰρ τὸ παράπαν ἔστι τῶν ὄντων οὗ μὴ παρὰ τῷ Θεῷ πάντως ὁ λόγος προένεστιν. Ὧν δὲ παρὰ τῷ Θεῷ προϋπάρχουσιν ὄντες τῆς οὐσίας οἱ λόγοι, τούτων δηλαδὴ κατὰ πρόθεσιν θείαν πάντως ἐστὶν ἡ γένεσις.*

[84] Perl (1991), 169. [85] Ibid. 161.

are not',[86] this seems incompatible with his previously stated view that the Logos is the highest universal containing 'particular *logoi*', such as *logoi* of created genera. The inconsistency is especially acute if '*logoi* of created genera' implies that the *logoi* themselves are generic in their own essence.

Eric Perl's idea of the Logos as highest universal is based on the text from the *Ambiguum* 41, which I have taken as my starting point in this section.[87] I do not think anything in the text contradicts the interpretation I have given above. As far as I can see, it rather confirms it. Maximus talks about *logoi of* the particular, and *logoi of* the universal and the generic. When he says 'the *logoi* of what is most universal and generic are held together by wisdom' (ὑπὸ τῆς σοφίας συνέχεσθαι), and Christ 'holds together the universals of beings by power of wisdom' (συνέχων τῇ δυνάμει τῆς σοφίας), 'embracing their complementary parts by the sagacity of His understanding' (περιέχων τῇ φρονήσει τῆς συνέσεως), it is reasonable to adduce that the wisdom and sagacity of God mentioned here are identical with the divine knowledge in *De char.* 4.4, 'His eternally pre-existing knowledge of beings'. This divine knowledge is, on my interpretation, to be identified with *the sum of all the logoi*. The divine knowledge, or wisdom and sagacity, is the same as all 'the pre-existing *logoi* of what has come into existence' (οἱ τῶν γεγονότων προϋφεστῶτας λόγοι),[88] that is to say with all the *principles* of beings. These beings are arranged—on the metaphysical and ontological basis of these *logoi*—as an *immanent* taxonomic system of genera, species, and individuals. Christ is *not* therefore the highest universal. Yet He holds together all of the *logoi* that are principles of universal and particular being (οὐσία). Christ as the Logos is not a universal at all, but is the personal divine center of all creation.

The Maximian understanding of the divine Ideas (*logoi*) as acts of will (θελήματα) is of the greatest importance, and distinguishes his thought markedly from the mainstream of non-Christian, Neoplatonic philosophy. Maximus' exemplarism is a doctrine of divine principles, and these become expressions of the divine will.

The immanent order which God, by His *logoi*, institutes in the cosmic organization is the subject of the next section.

[86] Ibid. 162. [87] *Amb.* 41, PG 91: 1313a–b.
[88] Cf. the discussion of *Cap. gnost.* 2.4 in Ch. 4 § 1

III. THE ORDERING OF ESSENTIAL BEING—EXPANSION AND CONTRACTION

According to St Maximus, God by His *logoi* creates the world, both invisible and visible beings, and what He creates is instituted *in concreto* as a taxonomic system. This arrangement is established within a procession-and-conversion-scheme by the double ontological 'movement' of constitution, expansion and contraction (διαστολή and συστολή). I shall now try to identify the internal characteristics of this system, its inner logic. The basic category of Maximus' system of the world is essence (οὐσία).[89] First, I will seek to determine the content of this category. Second, by the end of this section, I shall interpret one important text from the tenth *Ambiguum* that sheds some light on the double movement of expansion and contraction.

Essence

It is well known that οὐσία carries a multitude of different meanings in the philosophical and theological literature. A starting point for investigating the interesting intellectual background of Maximus' doctrine of essence is the greatest kinds (μέγιστα γένη) of Plato's *Sophist*. These are Being, Rest, Motion, the Other (Difference), and the Same (Identity).[90] Plotinus incorporates the greatest kinds into his discussion of the Intellect.[91] He stresses how sensible and intelligible beings are contained by their archetypes in the Intellect. According to Atkinson, the Platonic genera, 'once fitted into the schema of the Aristotelian self-thinking mind, are not Forms in the sense that they are included among the Intellect's objects; they rather illuminate the way in which Intellect engages in its activity'.[92] The Intellect is at rest and its 'is' is for ever, according to Plotinus. As it is in this condition it is identical with Being, because what thinks (i.e. the subject) and what is thought (i.e. the object) are identical. It is the same 'entity' which thinks and is thought. On the other hand, thinking requires

[89] Cf. *Amb.* 10, PG 91: 1177b ff. and *Ad Thal.* 48, CCSG 7: 341, 180 ff. According to Zizioulas (1985) the Greek Fathers hold the basic category to be hypostasis or personal being. I cannot see, however, that this is what is taught by these Fathers.

[90] Cf. *Sophist* 254d ff.

[91] *Enn.* 5.1.4.

[92] Plotinus, *Ennead V.1*, comm. and trans. M. Atkinson, 95–6.

otherness because without otherness there would not emerge any object of thought. Therefore, on the one hand there is sameness, on the other hand there is otherness. Now, as we saw in Chapter 2 § III, the Intellect is somehow the Maker of the cosmos, containing within its thinking-process the archetypes of everything. The Intellect as Being or Essence is understood as an activity in which it thinks itself as an intelligible cosmos. The archetypes are transmitted to lower reality by the *logoi*, through the process of double activity, which I have discussed in Chapter 2 § III and shall return to in Chapter 5 § II.

The Maximian doctrine of essence shows similarities with the Platonic doctrine presented above. In Maximus we find not only a doctrine of a taxonomic tree of beings, a doctrine of rest and movement in the institution of essential being and of *logoi* as creative principles, but also the doctrine of differentiation and identification immanent to the system of essence, as we will see below in this section. The difference is, however, that in Maximus all this belongs to created otherness and not to the divinity. When I speak of essence as a basic category of Maximus' thought, 'category' should be taken in the sense that it is the basic ontological distinction of created beings which distinguish them from God who transcends essence and who makes them to be 'something' and not 'nothing'. (I will return to this later.)

Von Balthasar has some interesting observations of the meaning of of essence in St Maximus.[93] According to von Balthasar, Maximus uses essence in these two senses:[94] (i) First, it denotes the highest ontological category of being, a category which includes all species and individuals ('Gattungen und Individuen'). According to von Balthasar, this is not to be understood as a pure universal, but rather as the real totality of the world ('die reale Totalität des Universums'). What von Balthasar means by 'a pure universal' I can only guess, but the context seems to indicate that he is thinking of something like an abstraction, i.e. a definition in the mind of a scientist. Essence in the first sense is, however, *not* of this kind. Von Balthasar finds his

[93] Cf. Thunberg (1995), 83; von Balthasar (1961), 213 ff.

[94] Von Balthasar refers to two 'senses', but what he calls two 'senses' are not two different concepts, rather they are two aspects of the same basic idea.

evidence in the *Ad Thalassium*, in a text where Maximus refers the individuals to the species, species to genera, and genera to essence in a manner that seems to indicate the actual arrangement of the world and not abstract universals.[95] (ii) Second, the term οὐσία has nearly the same meaning as φύσις, and denotes 'the particular nature by itself' ('die einzelne, particuläre Natur'), but not as isolated 'being for itself' ('Für-sich-sein'), rather as something which, when viewed in distinction from concrete individuals, connotes something common and specific, because all nature is something common.

In my view, the two basic aspects of the Maximian essence are (i) 'common nature' and (ii) 'particular nature'. This distinction is central to Maximus' world-system.

Common Nature

Essence as common nature has two distinguishing marks: (i) it has its 'location' in particular beings, i.e. it is immanent to them, and (ii) it collects particulars together in wholes, so as to constitute family-groups of specific and generic character. When Maximus refers individuals to species, species to genera, and genera to essence, this reminds one strongly of the Neoplatonic doctrines of how species exist in the Intellect, but according to Maximus, everything is referred to a category that neither pertains to *abstract* universals nor to universal kinds as the intelligible content of the divine intellect. Rather we find here the *immanent nature* of created beings that join creatures into *communities*. The system is instituted by divine activity through the *logoi*. The *logoi* are the principles on the divine level, while the essence is something created by these *logoi*.

The idea of essence as the common nature which joins beings together in community, plays in the background when Maximus in the first chapter of the *Mystagogia* points to the relationships between cause, parts, and whole.[96] The parts seem to be understood as particular *creatures*, the whole could be interpreted as the *common nature* joining the particulars together, and the cause is probably the *logos*. The cause (ἡ αἰτία) reveals the whole and the whole reveals its parts,

[95] *Ad Thal.* 48, CCSG 7: 341.180 ff.

[96] *Myst.* ch. 1, PG 91: 665a.

Maximus says, and 'by the cause the same whole, and the parts of the whole, come into *appearance* and *being*'. A number of particulars are not just an accidentally collected group of beings, but also a group with certain common features. What is important to understand here is that these common features are immanent in the particulars. The *logos* reveals a multitude of particulars not only as a *collection*, but also as a *whole* because these same particulars *qua* created are given a common essential determination by this *logos*. The particulars become a whole in the sense of being a species. As such a species they are coordinated with other species and grouped together under a genus, which again should be seen under a still higher genus in the immanent taxonomy of the created world.

It is worth noting that in the *Mystagogia* both the whole and the parts are given appearance and being by the same cause, i.e. by the *logos* of being. A *logos*, however, primarily seems to be the cause of particulars. It could therefore be claimed that the essence, both as common nature and as a universal, depends on the existence of the particulars.

The *logoi* are the divine principles for the world-system and *οὐσία* is distributed taxonomically in the created world. A text from the seventh *Ambiguum* should be considered in connection with this. Created beings, Maximus says, are comprised by their own *logoi* and by *logoi* external to them.[97] This text should be compared with the taxonomic system as presented in *Ambiguum* 41:[98] in the created world, particulars are contained by their species, species by their genera, genera by genera on higher levels of the system. In other words, in the immanent system different beings are *unified* on higher levels, and in the end the whole cosmic building is unified in the highest category of essence.[99] Another way to express this unification is to point to the principles of the system, as is done in *Ambiguum* 7. One could then say that created beings are comprised by their *logoi*, i.e. particulars by the *logoi* of their species, but in addition by the *logoi* of their genera etc. The whole idea could be illustrated by a set of circles as in Figure 3.

[97] *Amb.* 7, PG 91: 1081b. [98] *Amb.* 41, PG 91: 1312c ff.
[99] Cf. *Amb.* 10, PG 91: 1177b ff.

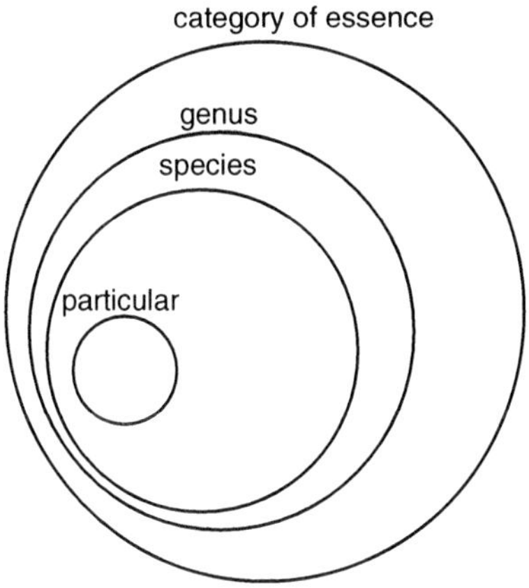

Figure 3. The unification of different beings in the immanent system

Let us explore further the concept of essence as common nature. In *Opusculum* 14 Maximus maintains that essence (*οὐσία*) is the same as nature (*φύσις*).[100] In *Opusculum* 23 he says that the essence is 'the form (*εἶδος*) itself and the nature (*φύσις*), i.e. that which exists by itself'.[101] In this way essence, nature, and form are used for the same concept. Essence and nature are further said to be *common* and *universal* (*κοινὸν καὶ καθόλου*), and since essence, as we saw, is both *form* and nature, we understand that these predicates also characterize the form.[102] These texts clearly establish the idea of 'common nature' as the one aspect of the Maximian concept of *οὐσία*. The introduction of the term nature is interesting: in earlier theologians 'nature' could be taken in the same sense as hypostasis, which means a concrete form of being. The council of Chalcedon, however, distinguished between nature and hypostasis and speaks of two natures in Christ.[103] Nature in the sense of *essence*, as something common and even universal, is primarily a real and *immanent* principle of community and universality, not an *abstract* universal.

Essence as an immanent principle of community should be seen as a *differentiated essence*. This concept makes possible the internal dynamics of the Maximian taxonomy of the world. To get a clearer

[100] *Th. pol.* 14, PG 91: 149b. [101] *Th. pol.* 23, PG 91: 260d–261a.
[102] *Th. pol.* 14, PG 91: 149b; cf. *Th. pol.* 23, 264a–b. [103] Tanner (1990), 86.

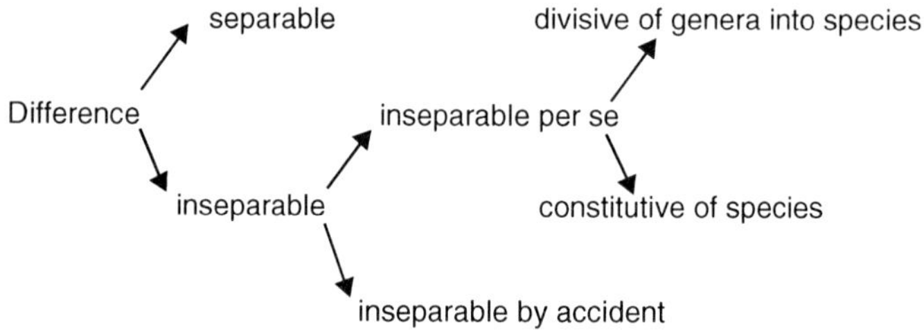

Figure 4. The Porphyrian division of the kinds of differences (*Isagoge*, CAG 4.9.10–10.3).

idea of differentiated essence and the dynamics of the system we must first study Maximus' doctrine of *difference*. We may, he says, contemplate 'the infinite difference of natural beings' (*τῇ φυσικῶν τῶν ὄντων ἀπείρῳ διαφορᾷ*), and then become aware of the *logos* according to which they were created.[104] A difference (*διαφορά*) between beings is, consequently, the effect of a *logos* of creation, and the contemplation of a given difference points to the *logos* from which it is instituted. In *Ambiguum* 22 Maximus says that things are different by reason of the *logoi* by which they subsist according to their essence.[105] A created being, then, has subsistence as an ontologically defined being, different from other beings, because of a differentiated essence established by a *logos*.

The term 'difference' is used by Maximus in a way which strongly reminds one of the 'form-making difference' (*εἰδοποιὸς διαφορά*) and the 'constitutive difference' (*διαφορὰ συστατική*) of Porphyry's *Isagoge*. We have difference in the strict sense, says Prophyry, when two things differ because of a form-making difference. A man differs from a horse because of such a difference that makes each being 'other' (*ἄλλο*) in relation to others.[106] According to Porphyry, the being of each thing is one and the same (*ἓν καὶ τὸ αὐτό*) as the result of such a form-making difference, which has a constitutive (*συστατική*) function for the essence of a being.[107] See Figure 4 for an illustration of the way Porphyry divides the kinds of differences.

[104] *Amb.* 7, PG 91: 1077c.
[105] *Amb.* 22, PG 91: 1256d.
[106] Cf. *Isagoge*, CAG 4: 8.15–9.6.
[107] Cf. *Isagoge*, CAG 4: 9.10–10.3.

In his use of the term, Maximus has probably learnt from several earlier thinkers, both Christian and non-Christian. The concept of difference played a major role in the logic and dialectic of Aristotle. The Neoplatonic commentators on the *Organon* were, of course, well aware of the concept. The famous *Isagoge* of Porphyry became important for Christian thinkers, theologians, and philosophers, in the East as in the West.[108] Many commentaries were made in the Neoplatonic schools, both on the *Categories* of Aristotle and on the *Isagoge*. From the Alexandrian school, with the works of which Maximus may have had some acquaintance, we have a commentary on the *Isagoge* by the Christian Neoplatonist David, and commentaries on the *Isagoge* and the *Categories* by Elias.[109] A philosophical source of primary importance could be the logical compendium or handbook which, according to one hypothesis, Maximus probably owned himself. It contains sections on the predicables and on the categories.[110]

Other Christian authors who used this terminology before Maximus include St Cyril of Alexandria, who employed the term in Christology; Dionysius the Areopagite, who gives it a cosmological sense; and Leontius of Byzantium, who seems to build directly on Porphyry.[111]

In Christology, Cyril of Alexandria had pointed to the existence of difference (διαφορά), but denied division (διαίρεσις) in Christ.[112] The Council of Chalcedon likewise denied that the difference between the natures was removed by their union in the one person or the one hypostasis of Christ, and affirmed that the property of both

[108] Cf. the *Capita philosophica* in *Dialectica*, first part of the *Pege gnoseos* by St John of Damascus. Cf. also the commentaries of Boethius to the *Isagoge*.

[109] Published in CAG 18, parts 1 and 2.

[110] Roueché (1974), 74 ff. Cf. § iv of the Introduction.

[111] Cf. Thunberg (1995), 51 ff. Thunberg gives an interesting analysis of St Maximus' use of the term. For the logic of St Cyril of Alexandria, cf. Siddals (1987).

[112] In *Scolia de incarnatione unigeniti* (PG 75: 1385c) St Cyril says that the one Lord Jesus Christ must not be divided (οὐ διοριστέον) in what is is characteristic of humanity and what is characteristic of divinity. Rather the differences of the natures must be safeguarded, and the natures must not be confused. A comparison with *Adversus Nestorium* (PG 76: 85a–b) shows that οὐ διοριστέον means that διαίρεσις is excluded. Cf. ὁ τῆς ἑνώσεως λόγος οὐκ ἀγνοεῖ μὲν τὴν διαφοράν. ἐξίστησι δὲ τὴν διαίρεσιν, 'the *logos* of unity is not unknown with difference, but drives away division'.

natures was preserved in the union.[113] The Christological use of the term is important, because it gives direction to Maximus' employment of it in his cosmology and his anthropology. The definition from Chalcedon provides him with a logical tool that makes its imprint on the whole of his thought about the relations between God and the world, and the internal relations of the created cosmos. This we may refer to as the 'Christological logic' or 'Chalcedonian logic' of Maximus. At the core of this logic are the famous four adverbs: *ἀσυγχύτως, ἀτρέπτως, ἀδιαιρέτως, ἀχωρίστως* (no confusion, no change, no division, no separation). In Chapter 5 we shall return to the role these concepts play in the thought of Maximus.

It is easy to see the Christological relevance of the following definition which Maximus gives of 'essential difference': 'a *logos* by which the essence, that is to say nature, remains both undiminished and unchanged, unmixed and unconfused' (*λόγος καθ' ὃν ἡ οὐσία, ἤγουν φύσις, ἀμείωτός τε καὶ ἄτρεπτος, ἄφυρτος ὁμοῦ καὶ ἀσύγχυτος διαμένει*).[114] It seems that here Maximus uses the concept of essential difference in a sense which is close to Porphyry's form-making and constitutive difference, which we discussed above. We should also note the important idea of the integrity of nature that is basic to Maximus' thought.

On the one hand, diairetic differences *divide* the genus, but, on the other hand, they function as *constitutive* on the level of the species. If man is a rational animal, then the difference 'rational', on the one hand, functions as divisive (*διαιρετική*) of the genus animal, and, on the other hand, as constitutive (*συστατική*) of the species man. As constitutive or essential, the differences are contemplated with a view to the community they establish between beings of the same species. The difference 'rational' is common to all individual men, establishing them as *one* species. Seen in this way, differences are mainly understood as collective or inclusive.

In Maximian doctrine, the internal play between the genus and the (dividing and constitutive) differences are dynamic relations in the real world. In the end everything is both differentiated and unified

[113] *οὐδαμοῦ τῆς τῶν φύσεων διαφορᾶς ἀνῃρημένης διὰ τὴν ἕνωσιν, σῳζομένης δὲ μᾶλλον τῆς ἰδιότητος ἑκατέρας φύσεως* (Tanner (1990), 86).

[114] *Th. pol.* 14, PG 91: 149d.

in the category of essence, which means that they are differentiated without separation and unified without confusion. In this way the category of essence is the basic structure in the whole system of created being.

This dialectic of difference and identity is seen most clearly in *Ambiguum* 41.[115] On every level where plurality is present, it is due to differences introduced by God into the system of cosmic organization. But differentiation does not prevail, because the differentiation between, for instance, species on the lower level, is 'released as it were from the variety caused by difference, and [the species] find identity one with another'.[116] In the technical vocabulary developed above, this means that species are identical in genus because of the generic essence that is in every one of them.

Naturally, Maximus would not accept a diairetic division that is non-dialectical and unilateral. If the division represented a given ontological fact in such a way that it reduced the fundamental structure of reality to the horizontal order, then it would follow that such a cosmos is not only naturally divided; in addition, each thing in it would be separated from every other, leaving it isolated in its own irreducible and monadic existence. The vertical dimension would then have to be considered as a theoretical and scientific construction of 'sortal-predicates' which would not establish any ontological community on a more basic level.[117] This would be almost a nominalist position. From a Maximian point of view, a sceptical attitude to division could be motivated by the insight that if reality were conceptualized as simply divided, this would entail the rejection of the basic relationship that each being bears towards every other. Such a rejection would in Maximus' eyes be morally reprehensible, and would deny the human task of mediation within the divinely instituted order of the cosmos. According to Maximus, a positive dialectic takes place in the taxonomic system of reality as God originally created it. What we find at work in the cosmic construction is the principle of unity-in-plurality.

[115] *Amb.* 41, PG 91: 1312c ff.

[116] *Amb.* 41, PG 91: 1312d, trans. Louth (1996).

[117] According to John Locke, the essence of a genus or a *sort* is nothing but an abstract idea for which a general or a *sortal* name stands. Cf. *An Essay concerning Human Undestanding*, 3: 3.15. Cf. Lloyd (1981), 1.

St Maximus' main point apropos the first aspect of essence is to demonstrate that beings in the world are somehow collected together in natural communities or societies comprising every particular of the different kinds. Eventually all beings demonstrate the same basic identity of being instances of essence (οὐσία), and this means that they have the common denomination of being *something* and not *nothing*. Every creature is grouped together with other creatures in the concrete composition of the cosmos, to be a plethora willed by God. Now, a world composed of such family groups would obviously not exist without particular beings. There have to be individual creatures that make room for *common natures* in their own being. The common nature of the particular is the principle of the societies of beings. This is a kind of *realist* position that claims that community between particulars springs from certain universal features immanent in the particulars themselves. From a philosophical point of view, is this a reasonable position? I believe it could be defended. Peter and Paul are individual men, but each betrays particular features that nevertheless bear sameness to the other. Each is an intelligent creature, and 'intelligent' shows a specific nature, common to both. This makes it reasonable to group them together as members of a natural kind. Further, the animality of these particulars, as members of a natural kind, has features in common with other natural kinds, so that different species of animals may be collected together in genera, being inclusive wholes that eventually comprise the whole animal kingdom. According to Maximus, this line of thought could be extended to include every substance in the existent world.

For Maximus, this kind of realist position is not a coincidence to his overall doctrine. It is connected with his soteriological perspective and could be taken as a philosophical requirement of it. In *Ambiguum* 41 he develops the idea of man as a microcosm. The microcosmic being of man could be demonstrated by pointing out how man is related taxonomically with all created beings. Man is the natural *terminus* in the process of creation. It is man, first and foremost, that is created in the image of the Creator, and he contains the richest amount of essential characteristics. The microcosmic being of man makes him the natural bond (σύνδεσμος) between all levels of being, and he is created just for this purpose: to actualize the created potential of his being to achieve a fully realized community between all creatures and their Creator. When man was created, the ontological conditions were established, and the practical task of unification remained to be fulfilled. But man failed. It is no coincidence that God the Logos, in His Incarnation, assumed human nature, since precisely this nature was designed to be the starting point for the actualization of the divine purpose. The One who is the centre of all the *logoi* assumed human nature, and, by assuming it, He assumed the microcosm He had created as a link between every existent being, sensible and intelligible. This idea is emphasized in the *Mystagogia*. God, Maximus says, by His Power, 'leads all beings to a common and unconfused identity (*ταυτότητα ... ἀσύγχυτον*) of movement and existence', and 'all things combine with all others in an unconfused way by the singular indissoluble relation to and protection of the one principle and cause'.[118] Now, what is expressed here, is the ontological condition for universal salvation, that is to say, a salvation of all created beings. Every created thing has its final goal in God. This idea of a movement from plurality to unconfused unity is described in the *Mystagogia* in connection with man's ecclesial existence:[119] numerous men, women, and children who differ from each other in many ways, are born into the Church and given *one* divine form and designation. Each converges with all the rest and joins together with them in one, simple and indivisible grace and power of faith, *so as to appear as one body formed of different members*. This one body is the Church, i.e. the body of Christ, because He, as the text has it, 'is all in all'.[120] In the *Ambiguum* 41 Maximus says that Christ recapitulates the all in Himself, 'showing that the whole creation exists as one, like another human being, completed by the gathering together of its parts one with another in itself'.[121]

The taxonomic system described by Maximus, then, establishes the fundamental ontological conditions which make possible the salvation and deification of all creation. After the Incarnation, human beings are not placed outside the regenerative work of Christ. They still have their task to perform, viz. as members of the Church, the

[118] *Myst.* PG. 91: 664d–665a. Berthold's translation (1985), Greek words inserted by me.

[119] *Myst.* PG 91: 665c–668a.

[120] *Myst.* PG 91: 668a, quoting St Paul, cf. Gal. 3:28; Col. 3:11.

[121] *Amb.* 41, PG 91: 1312a–b, trans. Louth (1996), 160.

body of Christ. We shall focus on the human path to reintegration and salvation in § IV.

Particular Nature

According to Wesche, in Leontius of Jerusalem we find the concept of a *particular nature*; for instance the particular human nature of this man Peter.[122] We have a similar doctrine in St Maximus. This particular nature is not totally closed off from the common nature of the species. It is somehow the specific being as modified into something particular and joined individually to a certain person. Textual confirmation of this concept is found in *Ambiguum* 41 (see my interpretation of the basic level of the taxonomy, Ch. 3 § II), and in *Ambiguum* 10 (see below, present section). We shall return to this concept of essence in § IV of this chapter.

The two aspects of *οὐσία* are not to be understood as wholly separate senses, but as two aspects of one and the same structure of being. The being of the creature points, so to speak, in opposite directions. On the one hand, essence is *common*, i.e. it groups a being together with other beings in species and genera. On the other hand, essence is particular, i.e. it belongs to or is assumed by a particular existing being.

One could now get the impression that we are faced with a continuum, but, obviously, there is a *leap* from universality as such to particularity and vice versa. Common nature and particular nature must be distinguished more sharply. I hold fast, however, to what I said above, that we are dealing with two aspects of one and the same basic structure of being, i.e. the essential determination of a creature. The particular humanity of Peter is not a 'kind', but belongs to him alone and is not in any way shared by other human beings. On the other hand, the Petrine *humanity* could be viewed in distinction from Peter, and then we move in the opposite direction, towards universality: Petrine *manhood* is an instance of *being human* (i.e. of the specific nature), and this again is an instance of *being an animal* (i.e. of the generic nature), etc.

[122] Wesche (1987), 80 ff.

Universals

Interestingly enough, St Maximus' position on essence as common nature comes very close to one of the senses which Aristotle himself gives to 'universal' (καθόλου) in the *Metaphysics.*[123] It is called universal because it contains many particulars. It is predicated of each of them, and each is a unity (e.g. man, horse, god). What defines the universal here is its *containing of many particulars* (πολλὰ περιέχον). However, the basis of this universal or 'wholeness' is the fact that several beings have the same essence: 'because they are all living things', Aristotle says. Now, here we have something similar to the Maximian way of thinking: individuals are referred to the species because the individuals have a certain immanent character. Further, a species is not an abstract, universal concept, but a concrete family-group that embraces many individuals.

In *Ambiguum* 10, St Maximus talks about universals in an interesting way. The whole context, in which he is commenting on divine Providence, should be taken into consideration.[124] The permanence, order, position, and movement of everything; the agreement of the parts with the wholes; the differences and unions of created essences; all of this witnesses to the Providence of the Creator, says Maximus. Providence is defined as the will of God, 'through which everything that is receives suitable direction'.[125] Divine Providence as the will of God is, obviously, related to the theory of *logoi* as principles of the cosmic order. Now, this Providence is exercised not only over individuals, but over *universals* as well, Maximus asserts. This would seem strange if these universals are divine Ideas: that is to say, are *identified* with the *logoi*. God, surely, does not make His own being the object of Providence? The universals, however, are not transcendent Ideas, but, as is seen clearly from the text, are immanent in the created order. What comes next is of major importance for the doctrine of the primacy of the particulars. The universals consist of particulars, Maximus says, and if the particulars perish, then the universals will perish with them.[126] Clearly, the *logos* itself of any created being cannot perish. What could this mean then, except (i) that *logoi* and

[123] Arist. *Metaph.* Δ, 26.1023^{b}29 ff.

[124] *Amb.* 10, PG 91: 1188d ff.

[125] *Amb.* 10, PG 91: 1189b.

[126] *Amb.* 10, PG 91: 1189c–d. Trans. Louth (1996), 145.

universals are distinguished, (ii) that the universals together with particulars are perishable things, and (iii) that the particulars furnish the basis on which universals exist.

Expansion and Contraction

There is an important, but difficult, text on the double movement of expansion and contraction in the tenth *Ambiguum*.[127] Maximus says that 'essence itself' is subject to expansion and contraction, but what is to be understood by 'essence itself'? The answer seems to be that it is the essence of all kinds of created beings, i.e. of those subject to generation and corruption, and generally of those moved by the *logos* and the mode (*λόγος* and *τρόπος*) in accordance with expansion and contraction. The distinction between *logos* and mode will be discussed in the next section.

By essence itself is meant then, the essence of all kinds of beings. The primary focus here obviously is on *the particular* which is the only kind of being typically subject to generation and corruption and to movement by *logos* and *tropos*. If this hypothesis is correct, then by 'essence itself' is meant the common nature of the particular as immanent principle of community. But then the question arises: in what way can essence itself, as the highest universality of being, be the common nature of the particulars. This, however, is answered by pointing to the double movement of expansion and contraction. Essence itself is moved (*κινεῖται*) from the most generic genus through the more generic genera to the species, yes, even as far as to 'the most specific species', Maximus says. This is the expansive movement—but this gives rise to several questions: (i) what is 'the most generic genus'? (ii) What does it mean that *essence* itself 'is moved'? (iii) What does Maximus have in mind by this *movement* through the genera to the species? (iv) What is the difference between the species and the most specific species?

(i) The most generic genus is *οὐσία*, the highest universality of being, as the basic fact of created beings, i.e. what makes them be 'something' and not just 'nothing'. A creature is called forth from

[127] *Amb.* 10, PG 91: 1177b–1180a.

nothingness and is made to be a 'this something'. (ii) To say that essence as the common nature 'is moved' must be a way of expressing the logic of the institution of creatures. It could mean one of two things: (*a*) that essential being is established by a downward process in which there is created a hierarchy consisting of a highest genus, intermediate genera, species, and the most specific species. I do not think, however, that this is what Maximus has in mind. He does not have a metaphysical doctrine of intermediate beings of a more and less universal kind 'between' God and the particular creatures. (*b*) It could mean that the particular creature, which is the primary object of the divine concern, is created, as the *locus* of an essential being which ontologically could be determined in a hierarchic way. In other words, the common nature is instituted hierarchically in the particular in such a way that the essence of the creature could be said to be given as a 'something' (i.e. it has a highest genus), as determined by the features of diverse intermediate kinds, by having a specific nature and, eventually, to be of the most specific nature. If this is a correct interpretation, it adds to the philosophical justification of the idea that each created being by its innate essential characteristics is essentially related to every other being in the cosmos. (iii) The *movement* of essence itself then, should be interpreted as the process of the institution of beings, but not as if the essence was literally moved from level to level in some kind of metaphysical hierarchy. The movement is the process of expansion. Essence is somehow 'expanded' or, in other words, 'distributed' or 'processed' to the basic level of particular creatures by being instituted hierarchically in the whole extent of them.

(iv) The institution of creatures as the object of the divine ordering of the cosmos, terminates in *the species* and, finally, in *the most specific species*. Now, what is the difference between these two kinds? It is commonplace in ancient thought to say, e.g. that man is a species (εἶδος), and what are 'below' the species would be the particulars comprised by it. There could be no level of universal or common being 'between' the species itself and its particular instantiations. If this is so, we seem to be left with two possiblities: (*a*) Maximus' species here is what we would normally call a genus, for instance the genus animal, while the most specific species are the kinds (– the species) of animal, i.e. man, horse, etc. (*b*) The species is the lowest degree of common being, for instance the human species, while the most specific species is the instantiation of this species in the particulars, in Peter, Paul, etc. Since the first interpretation (*a*) is in accordance with Porphyry, it is, perhaps, most safe to adopt it. On the other hand, the second interpretation (*b*) is not improbable either. If we opt for the second one, it follows that at the basic level we find the essence in the second sense outlined above (i.e. particular nature). In the next section we shall look into Maximus doctrine of *logoi* for individual beings, i.e. *logoi* instituting the being of the individual as its *particular* essence, for instance the essence of Peter. This essence of Peter must be understood as the essence of the particular *man* Peter, i.e. as the typically Petrine *manhood*.

The movement of expansion, then, is the ontological constitution of the manifold cosmos and this constitution is the permanent feature of the cosmic building. Beings are given as a manifold, but as a manifold of a certain order, i.e. they exist within the objective bonds of relations between particulars of different kinds. The process of expansion, Maximus says, terminates in a circumscription of essential being towards what is below (*τὸ εἶναι αὐτῆς πρὸς τὰ κάτω περιγράφουσα*), while the contractive movement (*συστολή*) terminates in a limitation ('definition') of essential being towards what is above (*πρὸς τὸ ἄνω τὸ εἶναι αὐτῆς ὁρίζουσα*).[128] The expression 'what is below' is a bit enigmatic. The limitation of essential being towards what is above, on the other hand, can only point to the basic difference between creatures and God. All beings are united in the contractive movement, and creaturely being is sharply distinguished from the divine being. At the other end of the scale, it is tempting to take the 'what is below' as referring to non-being. But non-being should not be seen as a 'realm' in its own right, at the farthest extreme from the divine being, as if there existed an original tension between two 'entities', God and non-being. The divine essence, Maximus says in the *De charitate*, has no opposite.[129] However, I believe that the enigma may be solved in the following manner: created being is, on the one hand, sharply distinguished from the divine being; on the other hand, it is kept by God within the limits of being. As long as God keeps a being within these limits, it does not risk being dissolved into nothingness, even

[128] *Amb.* 10, PG 91: 1177c. [129] *De char.* 3.28, PG 90: 1025b.

though the possibility of such dissolution is inherent in the creaturely status as such. This possibility is, in Maximus' view, no empty threat, even though what is created never will be completely annihilated because it is in fact circumscribed by divine Providence.[130] The internal dimension of possible non-being makes its claim on the *sinful* human condition after the fall, manifesting itself in internal disorders, passions, illness, and ultimately in spiritual and physical death. But the creature will never lose its divine gift of being, even though it only hangs on to it by a thin thread.

The idea of the movement of expansion is to show that beings are distributed as a *plurality* in an orderly fashion. The principles of this orderliness, on the other hand, point to the possibility of a *unification* of all creatures with each other. The contractive movement brings the mutual ontological relationships forth, and stresses the basic unity between beings. If one views the world as a downward movement one will end up with the plurality of beings. If one views it the other way round one realizes its fundamental unity. In the divine economy of the created world the double aspect of unity-in-plurality is preserved as a permanent, irreducible fact.

Even if much is different, this section on expansion and contraction shows similarities with Platonism. In Plato as in Plotinus Being (οὐσία) is connected with Motion and Rest. In Maximus, expansion and contraction are movements of Being and the possibility of rest lies in the fact that all beings are in communion in the structure of unity-in-plurality. Plato and Plotinus talk about the greatest kinds of Sameness and Otherness, and we have seen above the important role played by differentiation and identification in the Maximian system of essence.

Finally it should be said that the constructions of Porphyrian trees or taxonomies mostly serve as *examples* illustrative of the logical procedure of dividing a concept according to certain rules. If one believes that this kind of scheme is going to fulfil its intentions of mapping the whole compass of essential being, a great many presuppositions must be carefully worked out. It is easy enough to detect weak points in almost every example given to illustrate the Porphyrian tree. However, Maximus' idea of a real hierarchic relatedness of every being

[130] *Myst.* ch. 1, PG 91: 668a–b.

with every other in an ascending scale should at least be possible in principle, even though it is extremely difficult to *describe* theoretically the details of such a taxonomy.

IV. THE ONTOLOGICAL CONSTITUTION OF CREATED BEINGS

Before we inquire into the ontological constitution of created beings, we shall take a brief look at some basic categories of the Maximian world-system. According to St Maximus, every movement has an origin (ἀρχή), because every movement has a cause, and everything that has a cause has an origin.[131] Thus every created being in motion has an origin (moving cause) or *terminus a quo*, and an end, or *terminus ad quem*. Within the limits of these two *termini* the basic created conditions of created beings are the categories of time and space, as modern philosophers would term them.[132] Time and space are the given 'in-between', limited by God's efficient and final causality.

How are space and time employed by Maximus, and what meaning do they have in his philosophy? If we turn our attention to an interesting section of *Ambiguum* 10,[133] we find constantly repeated the terms (τὸ) ποῦ and (τὸ) πότε, meaning 'where(-ness)' and 'when(-ness)'. 'Whereness' is defined as something being *in a place*, and 'whenness' is defined as something being *in time*. Both these ways of being are qualified as 'to be in a certain way and not simply', because only God is simply, or, to be more precise, transcends 'where?', 'when?' and even being itself.[134] The expressions 'something being in place' and 'something being in time', are important. Some further comments must be added.

In Aristotelian philosophy, the category of 'where?' is not immediately equivalent with 'space' in the modern sense. It should rather be understood as the *place* of a given being. To point out something's place is no empty determination of it, but amounts to defining a basic characteristic of its being. According to Aristotle in the *De caelo*,

[131] *Amb.* 10, PG 91: 1177a. [132] Cf. *Amb.* 10, PG 91: 1180b ff.
[133] Ibid. [134] Ibid. 1180d.

the basic elements of the cosmos have *natural* movements that bring them to *natural* places.[135] For instance, earth moves naturally towards the centre of the cosmos, fire moves upwards, etc. Beings which are composed by the elements find their natural places in accordance with the amount of the given elements in their constitution. In the twelfth book of the *Metaphysics*, Aristotle distinguishes between three kinds of substances or essences: those that are both sensible and perishable; those that are both sensible and eternal; and the immovable substance.[136] Plants and animals are of the first kind, the celestial bodies of the second, and God the third. What we find here is a distinction between beings according to the place of their existence. Plants and animals are species that require bodies to fulfil the possibilities connected with their forms. Their natural place is in the sublunary sphere. What exists respectively in the sublunary and in the celestial places, are distinguished by certain properties that give them different degrees of ontological permanence and cosmic importance. God, however, is not composed of elements, but transcends materiality altogether. He occupies no place at all.

It is important to note that the designation of natural place is connected with the Aristotelian conception of *final causality*. Final causality is understood both horizontally and vertically. On the horizontal level, sensible beings strive for the realization of their forms, yet there must also exist cosmic conditions for this striving in the vertical order. The ultimate condition for all actualization in the cosmos, the universal final cause, is God (the first Unmoved Mover).

The theory of natural place and final causality was not weakened in the Hellenistic and the early Christian periods. The doctrine of final causality recommended itself to Christian thinkers because of their belief in divine providence.

According to Maximus, created beings, i.e. visible ones, exist in certain places at certain times. This fact is a primary ontological feature characterizing them. The division of everything created into *levels* of being in the opening section of *Ambiguum* 41 also testifies to this. The levels may be considered kinds of places. To be in time and place impose limits on created beings, which for this reason are neither without beginning nor uncircumscribed. A being having

135 Arist. *De caelo* book 1, ch. 8. 136 *Metaph.* Λ, 1.1069ᵃ30 ff.

things existing *before* it is not without origin, but its existence has a temporal beginning. This beginning is *from* the creative act as moving cause, and in addition, the being has a *terminus ad quem* towards which it moves. Moreover if a being has things existing *alongside* it, it is circumscribed by them and thus not uncircumscribed according to place.[137]

Time and place are created as the basic conditions of created beings within the causal limits set by divine efficient and final causality. Time and place exist simultaneously (ἅμα), and the one cannot exist without the other.[138] Within these conditions beings strive toward the natural goal for the entire universe, the final consummation of all creatures in the divine gift of deification. God is the 'from which' and the 'to which' surrounding the existential fact of the place–time condition.

To what degree are the Maximian definitions of time and place Aristotelian? According to Louth, Maximus takes his definition of place from Nemesius' book *On human nature.*[139] Nemesius says:[140] 'Place is the limit of the container by which it contains what is contained.' The second part of Maximus' definition is reminiscent of Nemesius' formula:[141] 'Place is the outside circumference of the all, either the position that is outside the all, or the limit of the container in which what is contained is contained.' It should be noted that place in this definition is not the place of a particular being; rather it designates the place of the whole cosmos. This makes no difference in principle, however, because the same definition will apply to the place of any particular being within the cosmos. When compared with Aristotles' definition of place in the *Physics*, Maximus' conception could be termed Aristotelian:[142] place is 'the containing body's limit which is in contact with what is contained'. Maximus' source is probably Nemesius, although his definition may also derive from several different Neoplatonist commentators on the *Physics*—or he may have constructed it himself.[143]

[137] *Amb.* 10, PG 91: 1181b. [138] Ibid. [139] Louth (1996), 209 n. 100.

[140] *De natura hominis*, PG 40: 600b. [141] *Amb.* 10, PG 91: 1180c.

[142] Arist. *Physics* book 4, 4.212^{a}6–6^a

[143] There are discussions of the nature of place for instance in Simplicius' commentary on Aristotles' *Physics* (*In Phys. 1–4* (Berlin 1882), CAG 9; Simplicius, *On Aristotle Physics 4.1–5, 10–14* (London 1992), and *Corollaries on Place and Time*

The text of the tenth *Ambiguum* contains too little information on the concept of time to evaluate its Aristotelian origin. But in *Cap. gnost.* 1.5 (PG 90: 1085a) we find that time is defined as the measure of movement and is circumscribed by number. This is essentially what Aristotle himself says in the *Physics*.[144] The immediate source once again could be some Neoplatonist commentator. The definitions of place and time with which we are concerned were common knowledge in the intellectual circles of the times, so that it may be quite futile to look for some particular source for Maximus exact wording.

'Where' and 'when' are generally thought to belong to Aristotle's famous categories of being. Maximus' theory of categories, however, differs in some respects from Aristotle's. The primary category, according to the Stagirite, is οὐσία (essence, substance). All other categories are related to substance as predicates of one or another kind. The condition of being qualified and quantified, 'where' and 'when' and so on, are dependent upon substance as the basic instance of being.[145] Yet according to Maximus, nothing can exist at all, 'neither essence, nor quantity, nor quality, nor relation, nor action, nor passion, nor movement, nor habit', in separation from 'where' and 'when'.[146] This list of categories (except for movement and habit [κίνησις and ἕξις][147]), contains genuine Aristotelian categories. The text of the *Categories* was one of the most popular to comment on in the Neoplatonist schools, so it is no surprise that Maximus considers such a set of categories as attributes or characteristics of being which delimit the cosmos. However, the basic role conceded to place and time is also found in St Gregory of Nyssa. According to Gregory, created beings are characterized by the fundamental fact of *extension*

(London 1992), both trans. J. O. Urmson). Simplicius defines place as 'the immediate immobile limit of the container' (584, 18). Cf. John Philoponus, who in addition attacks Aristotelian ideas of place (*In Phys. 4*, Berlin 1888, CAG 17; cf. Philoponus, *Corollaries on Place and Void* (London 1991), trans. D. Furley and C. Wildberg).

144 *Physics*, book 4, ch. 11.

145 Cf. Aristotle *Categories* 5.2^{b}3 ff., *Metaph.* Z, ch. 1.

146 *Amb.* 10, PG 91: 1181b.

147 It is a bit surprising that St Maximus adds passion, movement, and habit to the list of categories if his source was the logical handbook described by Roueché (1974), 74–5. These three categories are not discussed this handbook, even though there is a section on τὸ ἔχειν. Could it be that ἕξις has taken the place of τὸ ἔχειν in St Maximus? In the *Categories* ἕξις is included under quality, cf. ch. 8. Of course, St Maximus may have had access to sources in which 'habit' has taken the place of 'having'.

(διάστημα or διάστασις).[148] Von Balthasar describes this fundamental fact as 'that "receptacle" of all material being which God created in the beginning... and which is time and place'.[149] In this conception of place and time there is a tendency to make these predicates fundamental, i.e. to emancipate them from dependency on substance. This tendency was particularly emphasized by John Philoponus with respect to the concept of place. It seems that the concepts of place and time in Maximus are developed along the same lines. Yet he does not consider place and time to be some kind of substantial beings in their own right; rather they are created as the *conditions of created being*, and within the limits of these conditions, the primary intention of the divine creative act manifests itself in the institution of natures or essences. Place and time, therefore, are related to essence in the creation of the world. From an ontological point of view they are of equal primordiality (ἅμα).

The divine causality is manifested through the *logoi*. In *Cap. gnost.* 1.10 (PG 90: 1085d–1088a), Maximus says that God is the beginning, middle and end (ἀρχή, μεσότης, τέλος) of every created being.[150] These three terms are indicative of a threefold causal pattern: 'For He is beginning as Creator, middle as Provider, and end as Goal, for it is said, *from Him and through Him and for Him are all beings*.' The *beginning* or *origin* then, points to God as efficient-formal cause of being, while the *middle* indicates that God is the preserving or sustaining cause. The *end*, which is the third causal moment, points to God as the final cause of beings.

The ontological analysis of being in the first ten chapters of the first part ('century') of *Cap. gnost.* is developed further according to a pattern of triadic structures. The triad of origin, middle, and end not only describe the external metaphysical conditions of every being, but also indicate an ontological fact immanent to the creature itself. The creature is understood in accordance with these two triads:[151]

148 Cf. von Balthasar (1995), part I, ch. 1, with many references to the texts of St Gregory. The French word for διάστημα and διάστασις as translated by von Balthasar himself, is 'espacement', the translator into English has coined the word 'spacing'. I prefer 'extension' because 'spacing' most properly refers to space only. Cf. the important discussion of extension in Sorabji (1988), part I.

149 Von Balthasar (1995), 29–30.

150 According to Neoplatonic thought, this triad defines the completeness of being. Cf. Philoponus, ed. Wildberg (1987), 51 f.

151 *Cap. gnost.* 1.3, PG 90: 1084a–b.

beginning (origin)—middle—end
essence—potentiality (power)—activity (actuality)

The triads are intimately connected: an essence is the *origin* of a potentiality, that in relation to essence and actuality constitutes a *middle*. This potentiality has its *end* or *consummation* in actuality. According to Maximus the triad essence–potentiality–actuality is typical of 'all being'. It seems to me, however, that the triad is primarily designed to define rational creatures. As we shall soon see, potentiality lends itself to highlight the phenomenon of a natural will and is related to the subject of freedom. Even though some higher animals could be said to enjoy a certain amount of will and freedom, it is man that first and foremost has these characteristics.

The potentiality, determined both in relation to the essence and in relation to the actuality, has two aspects. As related to the essence it is *the capacity of a being to suffer a change in itself*. Then, in relation to the actuality, Maximus implies a further aspect, viz. that this capacity to suffer a change in itself in addition is *a capacity to enter a certain condition*. When a being effects such a change through its twofold potentiality, then activity originates as a certain 'dynamic' condition. In this condition the being achieves its goal, its consummation (actuality). What we are touching at here is the ontological structure of the phenomenon of will, a topic to which we shall return below.

In this connection it is also interesting to note that Maximus agrees with Aristotle with regard to the last-mentioned aspect of potentiality, which implies a passing-over to actuality. Aristotle is of the opinion that when a being, on the basis of its potentiality, actualizes itself in an *action* (πρᾶξις), then this action can be considered to include the end in itself.[152] According to Maximus, however, this 'natural' actualization does not reach its final consummation within the framework of a pure natural potential internal to a being, but has a further end.

What I have in mind is, of course, the doctrine of *deification*.[153] When I say that the natural actualization is not finally consummated within the limits of a natural potential, this could seem to imply a contradiction. I think, however, that deification is somehow

[152] Cf. *Metaph*. Δ, 6.1048^{b}18 ff.

[153] For the doctrine of deification, cf. Thunberg (1995), 427 ff., and esp. Larchet (1994).

a 'natural' actualization or consummation of a being, but that this 'natural' consummation is not within the power of the created being as such. Thunberg says that deified man 'suffers an ecstasy which brings him *outside himself*, i.e. beyond his natural capacity, in pure grace'.[154] I would add that deification is 'natural' in so far as it is the fulfilment of the divine intention of creation. The *logos* of a being could be termed its 'natural' *logos*, since it defines and constitutes its nature. This *logos* of being should not be isolated from the other *logoi* relevant to a given being, such as its *logos* of eternal well being. A typical Maximian expression of this complex may be found for instance in *Ambiguum* 7:[155] 'And he [i.e. man] is a "part of God", as *being*, because of his *logos* of being in God, as *good*, because of his *logos* of well-being in God, and as *God*, because of his *logos* of eternal being in God, to the degree that he has honoured these and been active in accordance with them.' Later in this section and in Chapter 4 § IV I shall consider the relation between these *logoi*, and ask whether they really are different *logoi* or aspects of one *logos*.

When I say that deification is natural, all I mean therefore, is that a being is created with a natural *potential* for deification.[156] In my opinion this much is implied in all expressions about moving naturally, because such movements are an actualization of such a potential.[157] But this potential does not imply that deification is achievable by the efforts of created being as such, rather it is a gift from God. As we shall see in the next chapter, when the creature is actualized in a good way according to its immanent potential, then it is graciously brought into what Maximus would call a new *mode* (*τρόπος*) of being.

An essence, Maximus says, has in itself a limit (*ὅρος*),[158] and this statement is important for the question of the ontological constitution of created beings. The limit may be understood as an *essential limitation* or *determination*. In one of the logical texts published by Roueché, a limit is described not as an abstract definition expressed in language, but as an ontological fact, a limitation in the thing itself, namely its generic and differentiated being.[159] Because of this determination an essence is the source of a certain kind of movement

[154] Thunberg (1995), 423. [155] *Amb.* 7, PG 91: 1084b–c.
[156] This will be commented on in Ch. 5 § IV as well.
[157] Cf. e.g. *Amb.* 41, PG 91: 1308c ff.
[158] *Cap. gnost.* 1.3. [159] Cf. Roueché (1994), 70 ff. 5 ff. Cf. Ch. 1 § IV.

(κίνησις) in accordance with a delimited potentiality. The movement is called an 'essential movement'. It is not this essential movement as such which is called 'actualization' above, but the essential movement culminates somehow in actualization as a final goal. With the term 'essential movement' we are given a further determination of *potentiality*, to which we shall return below. On this background it is quite clear that potentiality as such is not just a pure passive capacity, a 'not yet' in relation to a certain activity, a non-being before being, a privation before presence. By contrast, potentiality is something 'dynamic', a functioning capacity—all of which contributes to the impression that there exists a natural aptitude and potential of a being for its final deification.

Maximus says that the end of the essential movement is the actualized or active condition which is 'circumscribed naturally by its own *logos*'.[160] The limitation (ὅρος) which an essence has in itself, a limitation that determines its potentiality and circumscribes its activity, is due to *the presence of a logos*. These statements seem to contradict my thesis that the final goal of a being consists in transcending its natural capacity, but I do not believe it necessary to interpret them in such a way. Rather these statements can be taken as a confirmation of my interpretation. Since the consummation of a being is circumscribed by its *logos*, this implies that what is possible for a being is what is prescribed by its *logos* or *logoi*. This means that a 'logical' development of the created being from the natural actualization to its 'natural' *transnatural* deification by grace is possible. This must be what is implied in the scheme of *logos* of being, *logos* of well-being and *logos* of eternal being.

The interpretation of the triadic structure of created beings can be developed a bit further. In *Cap. gnost.* 1.5 (PG 90: 1085a), the triad beginning–middle–end is connected with *time* as the basic extension which is characteristic of such beings. This indicates that the being of creatures is somehow temporally extended according to the scheme of beginning–middle–end. Beginning indicates *past*, middle the *present*, and end the *future*. We may therefore, even though it might seem a bit speculative, take the liberty to supplement the two triads with a third one, which gives rise to the following schema:

160 *Cap. gnost.* 1.3.

beginning—middle—end
essence—potentiality—activity
was—is—will be

What further implications does this scheme have? What we find here is a threefold immanent structure of a created being, but this immanence is not locked up in itself. The past tense relates the being to the transcendent Beginning,[161] and the future to the eschatological End. The essence as immanent origin is therefore to be understood as an expression of and as related to its transcendent cause, which is the divine *logos* of its being. The actuality as immanent end is likewise related to the final end, which is the consummation of a being in deification, in accordance with the *logos* of eternal being. The potentiality is the 'middle' as the here and now, preserved by God, and points to the possibilities of free movement in the present. Still another triad may be brought forward, from *Ambiguum* 17, a triad that Maximus connects with the ideas of Beginning and End, and with the internal structure of beings.[162] This is the revised and anti-Origenistic triad of creation–movement–rest (*γένεσις–κίνησις–στάσις*).[163] The creative act is the *γένεσις* of created essence. Created essence is to realize its *movement* to end up in the *στάσις* which is the consummation of creation. The picture may therefore be further developed:

God
Beginning—Middle—End
logos of being—*logos* of well-being—*logos* of eternal being

the creature
creation—movement—rest

beginning—middle—end
essence—potentiality—activity
was—is—will be

[161] A. Louth (1996), 209 n. 101, makes some highly interesting observations about the use of the imperfect tense in *Ambiguum* 10.
[162] Cf. *Amb.* 17, PG 91: 1217c–d.
[163] Cf. Sherwood (1955), ch. 1, and 109 ff.

The three *logoi* here do not have to be three separate *logoi*; they could rather be three aspects of one and the same *logos*, or at least constituents of a closely interconnected triad. It is, however, difficult to argue this conclusively from specific texts. In *Ambiguum* 42 Maximus says that God united the *logos* of my being with the *logos* of my well-being, and healed the separation between these which I had caused. Through these two *logoi* God attracts me towards the *logos* of eternal being.[164] This seems to mean that the three *logoi* are not ontologically separated, but in so far as a separation occurs, it is in the creature's relation to them, and the creature is accountable for making a practical split in what should or even could not be separated actually. Maximus' doctrine implies that the *logos* of being has to do with the natural, immanent structure of being and the potentiality which something has received as God's creature. The *logos* of well-being concerns the voluntary actualization of this nature in God-willed projects. The *logos* of eternal being has to do with the fulfilment of the creature and the experience of deification. Seen this way, nature and deifying grace are distinguished, but not separated. There is so to speak a 'natural' connection between natural and transnatural, guaranteed by the triad of *logoi*.

The preconditioning essence makes present a potentiality (cf. the 'is') which is to be actualized (cf. the 'will be'). We have seen that potentiality is understood to be a moment of essential movement. This movement is typically creaturely, and takes place in the present within a range of possibilities for different concrete acts.[165] I understand the essential movement as the ontological structure of *natural will*. This is a major subject in Maximus' thought, concerning the details of which the reader should consult other studies.[166] Man is *essentially* (οὐσιωδῶς) willing, Maximus says in his disputation with Pyrrhus.[167] Maximus' idea seems to be that if the human essence consists in being a rational animal, and if man *qua* rational animal has a potentiality for movement, then this ability to move according to reason must be typical of man. The ability to move according

[164] *Amb.* 42, PG 91: 1348d.

[165] That movement is typical for created beings is said in many texts, cf. for instance *Cap. gnost.* 1.11, PG 90: 1088a.

[166] For instance Thunberg (1995), 208 ff.; Farrell (1989).

[167] Cf. *Pyrrh.*, PG 91: 301c.

to reason is understood as an ability of 'self-determinative motion' (αὐτεξούσιος κίνησις).[168] Here Maximus is in line with St Gregory of Nyssa, to whom the power of self-determination is integrated into the divine image in man.[169]

The potentiality to move by rational will into an actualized condition includes, as just mentioned, a *possibility* of choosing between different options. The will, identified as self-determination, is clearly understood as *free*. Now, is this power of self-determination similar to the autonomy of the will in the Kantian sense? The answer is undoubtedly negative. The 'self' (αὐτός) for St Maximus is a genuine self by the fact of its being constituted by the divine *logos* that sets the norm of its behaviour in accordance with the divine will. The self is an authentic self, therefore, only to the degree that it lets itself be moved in accordance with the divine intention. Genuine autonomy, consequently, is—if the term is allowed—strictly speaking 'theonomy'.

From the idea of self-determination Maximus concludes that if man is rational by nature, and what is rational is self-determinative as well, than man has by nature a will (θέλησις).[170] Consequently, an ability to decide how to move (i.e. to act) is an essential property of man, that is his natural will, which culminates in an activity or an actualized condition. For these reasons Maximus urges, apropos of the monotheletistic controversy, that the human nature which Christ assumed has a rational soul and a natural will, for 'everything that is rational by nature, certainly also is volitional by nature'.[171]

Because of this essential capacity for willing, it is natural for man to will in accordance with nature or the natural *logos*. The range of possibilities within which the essential movement takes place, is a range of good possibilities, with a view to which the creature is not determined in advance. This is the structure of natural freedom, which is not only a freedom from sin, but also a freedom to choose between diverse goods without a struggle of choice. Struggling over decisions is characteristic of fallen man.

[168] *Pyrrh.*, PG 91: 301b.

[169] *De hominis opificio* 4, PG 44: 136b–c. The same idea is found in Origen, cf. *De principiis* 3.1.

[170] *Pyrrh.*, PG 91: 304c.

[171] *Th. pol.* 7; PG 91: 77b.

Maximus has an interesting doctrine of virtue as *natural*.[172] Here he seems to be in disagreement with Aristotle who, in the *Nicomachean Ethics*, says, 'none of the ethical virtues arises in us by nature'.[173] The reason given is that what is natural has a certain necessity and cannot be changed by habituation. In the background here is the experience that men may reform their ways and become better by habituation. Virtues, Aristotle says, are acquired as the result of prior activities, i.e. just as we become builders by building and lyre-players by playing the lyre, we become just by doing what is just.[174] Consequently, 'virtues arise in us neither by nature (*φύσει*) nor contrary to nature (*παρὰ φύσιν*); but by our nature we can receive them and perfect them by habituation'.[175] In Aristotelian thought, being and goodness are intimately connected, even though Aristotle obviously makes a distinction between being and *moral* goodness.[176] Maximus' opponent in the *Disputation with Pyrrhus*, the former patriarch Pyrrhus, wonders how it is that virtues do not exist in all men equally, since we all are of the same nature.[177] Maximus holds that the virtues exist equally in all men because of the identical nature, and the reason why men differ widely in virtuous practice is that we do not act according to what is natural to us in an equal degree.[178] This makes room for ascetical practice in analogy with the Aristotelian habituation, but not in order to internalize virtuous patterns from the 'outside', rather in order to ward off deception, Maximus says.[179] In the sinful condition we are deceived as to our natural capacity. We don't know our true selves any more, and at least one step on the way according to nature (*κατὰ φύσιν*) is to acquire a renewal of the rational faculty.[180] Maximus could not have accepted a clear-cut distinction between being and goodness, or between ontological and moral goodness. The reason lies in the Maximian understanding of essence or nature. The essence is good as created by God, and all actions stemming

[172] *Pyrrh.*, PG 91: 309b. [173] *Nicomachean Ethics* book 2, 1.1103ª19.
[174] Ibid. 1103ª31 ff. [175] Ibid. 1103ª23–6.
[176] In the *Metaphysics* (A, 3.983ª31–2) the good is identified with the final cause. Cf. also the opening words of the *Nicomachean Ethics* and book 1, 4.1096ª23 ff. That Aristotle makes a distinction between being and moral goodness is shown in the section from the *Nicomachean Ethics* commented on above, the essence of man is one thing, his morality is another.
[177] *Pyrrh.*, PG 91: 369b. [178] Ibid. [179] *Pyrrh.*, PG 91: 309c–312a.
[180] Cf. *Amb.* 10, PG 91: 1109a ff.

from the essence are good, as long as they are in accordance with the *logos* of nature. They are good, not only in an ontological or generic sense, but morally so. The natural potential of intelligent beings is by definition a potential for virtuous acts. In the background lies, of course, the idea that nature is what God intends it to be. Nature is what is instituted by the *logos* of being, and the natural obligation of intelligent beings is not to separate in practical life the *logos* of being from the *logos* of well-being, with which it is connected in the divine scheme of things.[181]

Maximus' teaching on virtue as natural is in agreement with St Gregory of Nyssa's doctrine of the divine image in man. According to Gregory, God has introduced virtue into the original human nature.[182] A fundamental idea in both thinkers seems to be that there does not exist any ontologically pure nature to which must be added a supernatural grace in order for it to commit virtuous acts. Nature, as it comes forth in the creative act, is already graciously endowed with dispositions for the good. This is a natural condition in created beings, and in this natural condition there is already given an original and intrinsic relation to God. The natural will is the ontological condition which makes it possible for the creature to choose between alternatives, i.e. to actualize two basic possibilities, viz. deification or a fall away from God. To achieve one or the other of these possibilities the creature has to be movable in its very being. If it was not movable, then it could never sin nor—for that matter—be deified, because deification presupposes a voluntary cooperation between God and created being. The fall away from the community with God is unnatural, and the life of human beings separated from God is against nature. The way back from vice to virtue is, in effect, a way back to oneself, i.e. to the genuine human nature which God has ordained through His *logos* of human being.[183] We shall return to this in Chapter 4 § IV, because a further development of these topics should be made in connection with Maximus' doctrine of the three stages of spiritual development.

To clarify the ontological constitution of creatures a bit further, we must move into the sphere of the particulars in order to grasp how

[181] Cf. *Amb.* 10, PG 91: 1116a ff.

[182] *De hominis opificio* 4, PG 44: 136c–d.

[183] *Myst.* 5, PG 91: 676b.

Maximus understands the concreteness of being. In this connection we shall focus on the distinctive marks of particular being and try to determine whether Maximus tries to identify any principle or principles of individuation at work in the constitution of creatures. Man will be taken as the point of departure, but what may be discovered in the case of the human being could, with some consideration, be applied in the case of other creatures as well.

A short historical sketch may shed some light on Maximus' idea of particulars. In the *Categories* Aristotle defines primary *οὐσία* as individual (*ἄτομον*) and as numerically one (*ἕν ἀριθμῷ*).[184] It has been objected, however, that to be numerically one is not a property of primary substance only, because even the species could be said to have this *proprium*.[185] A species is, of course, a unity, but, in my opinion, the reason why they are not considered as numerically one in the *Categories* is that they are terms predicable on several particulars. In this regard they are somehow divisible. The particular, on the other hand, is not divisible, because it is not a predicable, and the only way to divide sensible particulars is to destroy them physically. It is in the *Metaphysics*, however, we find the famous Aristotelian text on individuation:[186] 'And when we have the whole, such and such a form in this flesh and these bones, this is Callias or Socrates; and they are different in virtue of their matter (for that is different), but the same in form; for their form is indivisible.' I believe it correct (as is traditionally supposed) that Aristotle, like Thomas Aquinas, held matter to be a principle of individuation.[187] But is it possible to explain what a genuine individual is by matter alone?

A particular being, according to Aristotle, is a whole consisting of form and matter. If we follow his doctrine of particular properties

[184] *Cat.* 3^b12–13. [185] Turcescu (1997), 67.

[186] *Metaphysics* book Z, ch. 8, 1034^a5–8: *τὸ δ' ἅπαν ἤδη, τὸ τοιόνδε εἶδος ἐν ταῖσδε ταῖς σαρξὶ καὶ ὀστοῖς, Καλλίας καὶ Σωκράτης. καὶ ἕτερον μὲν διὰ τὴν ὕλην (ἑτέρα γάρ), ταὐτὸ δὲ τῷ εἴδει (ἄτομον γὰρ τὸ εἶδος).*

[187] Several modern interpreters of Aristotle have argued that the Aristotelian form-in-matter is individual by nature. I can understand the need felt by modern philosophers to develop this interpretation, but I do not believe that there is sufficient evidence to claim that Aristotle himself in fact held such a doctrine. The doctrine of form and matter was developed to serve two ends. On the one hand, it was meant to explain how substances are ontologically constituted, and on the other hand, it should make it possible to grasp the essential contents of a substance. Cf. my article (in Norwegian) 'Aristoteles om *οὐσία*', *Opuscula* no. 1 (Oslo 1992).

in the *Categories*,[188] it would seem that the principle that makes properties into particulars is matter. In this respect it is the cause of the concrete presence of a substance in the world.

The Stoics made a distinction between primary and secondary subject or substrate (ὑποκείμενον).[189] The primary substrate is abstractly understood as unqualified matter. Substrate in the secondary sense is anything that is (a) commonly or (b) peculiarly qualified, in the way that 'the bronze, and Socrates, are substrate to whatever comes about in them or is predicated of them'.[190] Bronze, I suppose, is an example of something that is 'commonly qualified'. It is the substrate of, for instance, a statue. An individual being like Socrates, on the other hand, is defined as something 'peculiarly qualified', so that it is the combination of qualities that make this particular different from that.[191]

This idea seems somehow related to the concept of particular being which emerges within the Platonic tradition. In the *Theaetetus* 157b–c Plato suggests in passing that a sensible particular is a bundle (ἄθροισμα) of properties. This doctrine is later found in Antiochus of Ascalon and Alcinous.[192] According to Turcescu, however, it is Plotinus 'who took Plato's suggestion a little further and Porphyry who presented it in a more acceptable form'.[193] Turcescu asks the important question of what keeps the bundles of properties together. The answer, on Turcescu's view, is that, according to Plotinus, it is the soul and not some kind of material subject that carries out this unifying function.[194]

In Porphyry's *Isagoge* we find a description of the bundle-theory. He says something is called an individual if it is a unique collection of properties which in themselves are not unique.[195]

In the Neoplatonist Dexippus, a doctrine similar to the Stoic doctrine of individuals turns up again. But according to Dexippus, the problem is not to be solved by a concept of 'peculiar qualification' (τὸ ἰδίως ποιόν). This implies that in his view it is not by a conjunction of qualities, such as hooked-nosed, fair-haired, etc., that things

188 Cf. *Cat.* 1ᵃ23 ff. 189 Long and Sedley (1992), 1: 172.
190 Ibid. 1: 172 and 168, text E (Greek text, 2: 172, text E).
191 Ibid., cf. Turcescu (1997), 70–1. 192 Turcescu (1997), 74.
193 Ibid. 194 Ibid. 75. The answer is based on the whole of *Enn.* 6.3.
195 *Isagoge*, CAG 4: 7.19 ff. Cf. Turcescu (1997), 75.

primarily differ from each other.[196] What makes objects different according to Dexippus, is their *countability*. They are counted one by one and are therefore given as spatially separate. Dexippus does not, as far as I can see, deny that individual substances are peculiarly qualified objects, but he contends that the numerical difference, which makes them countable, to a greater degree establishes them as distinct items. Dexippus' idea of 'distinct items as countable' (διεστηκότα ὡς ἀριθμητά),[197] seems to me to imply a spatial model which could indicate that matter still plays a role in individuation since it localizes beings, even though he does not explicitly say so.

As far as I can see, what is typical of the Aristotelian and the Stoic doctrine is that particulars have a subject which somehow carries the properties. The bundle-theory, on the other hand, does not seem to presuppose any kind of material substrate. It is unclear what exactly is Dexippus' position on this. According to Turcescu, the Porphyrian definition of a particular is perhaps the most elaborate in the fourth century before the Cappadocians.

In St Basil and St Gregory of Nyssa we find the idea of hypostasis expressed in the term 'essence with properties' (οὐσία μετὰ ἰδιωμάτων).[198] The essence is understood as common nature and the properties as particularizing characteristics. We shall see below that Maximus comments on this definition, without, however, ascribing it to Basil or Gregory.

A bundle-theory is also found in Cappadocian thought, in Gregory of Nyssa. In two texts he explains his concept of material being in terms of such a theory.[199] The bundle-theory seems to emerge in his *To his Brother Peter, on the Difference between Ousia and Hypostasis* as well, when he defines hypostasis as 'the concourse of the peculiar characteristics of each [person] (τὴν συνδρομὴν τῶν περὶ ἕκαστον ἰδιωμάτων)'.[200] Maximus refers to this doctrine too, again without attributing it to Gregory.

[196] Cf. Dexippus, *On Aristotle Categories*, trans. Dillon, 62–3; CAG 4: 30.23 ff.

[197] Cf. CAG 4: 30.31–2.

[198] Cf. Turcescu (1997) and Grillmeier (1975), 372 ff. for a fuller treatment. Cf. St Basil, *Letter 236*, 401–2 (Loeb), St Gregory of Nyssa: *To his Brother Peter, on the Difference between Ousia and Hypostasis* (= St Basil's *Letter 38*, 197 ff. (Loeb)).

[199] *De anima et resurrectione*, PG 46: 124b–d; *In hexaemeron*, PG 44: 69b–c.

[200] Turcescu (1997), 73, cf. Pseudo-Basil, *Letter 38*, 218 (Loeb).

Now, the influence of the Cappadocian Fathers on later developments was enormous. Their concept of particular being returns in several later authors, even in the conceptually refined debates following the council of Chalcedon (451) and in sixth-century Byzantium.[201] The inadequacy of the definitions is felt in later literature, and the status of the Christological controversy required a refinement of the ontology to be applied to the clarification of dogmatic positions. I venture the hypothesis that in Maximus, if not earlier, e.g. in Leontius of Jerusalem,[202] some steps are taken in the direction of developing a more suitable ontology. I do not claim these steps to be revolutionary or very perspicuous, but they seem to indicate a further development.

Modern authors have claimed that Orthodox anthropology, based on patristic sources, has a distinctive doctrine of the human *person* to be distinguished from the typically western concept of an individual. The note is struck in the influential writer Lossky, when he says:[203] 'Purged from its Aristotelian content, the theological notion of hypostasis in the thought of the eastern Fathers means not so much *individual* as *person*, in the modern sense of this word.' Zizioulas gives a lot stronger expression to this belief.[204] On the one hand, it could be easy to embrace this theory because one could feel it would serve Trinitarian theology well. How could, for instance, St Gregory of Nyssa be rescued from tritheism if hypostasis and individual were not different things? On the other hand, it is a misunderstanding of the Cappadocians to believe that they tried to work out a philosophically adequate Trinitarian terminology. Rather they had the idea of developing pragmatic and metaphorical strategies for the usage of terms of what cannot be adequately grasped by the human mind.

For several years I have felt uneasy with the notion of person as clearly distinct from individual, especially since I have studied the sources without finding what is claimed to be taught in them. Törönen points out that all the characteristics commonly linked with a person in the modern conception—rationality, freedom, relatedness, and self-consciousness—in the Maximian context belong to

[201] One should consult Grillmeier (1995) on this.

[202] Cf. the interesting article by Wesche (1987).

[203] Lossky (1973), 53.

[204] Cf. Zizioulas (1985).

the nature of the species.[205] Törönen sums up Maximus' position, typical of the Greek Fathers, thus:[206] 'That, again, which is particular to one individual being, something which it does not share with other members of its kind, is what characterizes the hypostasis or person; therefore, particular, individual, personal, and hypostatic go together.' In the main lines I agree with this, but still I think there is a certain development in Maximus, compared with earlier thinkers. As a matter of fact there lurks a problem in Törönen's argument. If his treatment of hypostasis is kept in mind under his exposition on Christology, the composite hypostasis of Christ seems to split up into two hypostases, and I cannot believe Maximus would have exposed himself that obviously to the charge of Nestorianism. Törönen says the eternal Logos, who is a complete, simple hypostasis with a concrete and particular being, assumes in the Incarnation another form of being:[207] 'And he is particular also at this level.' I agree this has to be the case, but Törönen has commited himself to the view that for a nature to be manifested as concrete and particular is the same as being a hypostasis. If Törönen's point is that Christ has a complete human nature (i.e. a specific nature)—including body, soul, and mind—that exists particularized in a collection of concrete properties, what is it that does not make this into a hypostasis of its own—given the equation of particular with hypostasis? Some further consideration is—as far as I can see—required. Somehow there should be a hypostatic principle in Christ that would make His divine hypostasis the ruling principle or the ontological centre of His composite hypostasis. In that case, could we just equate particular and hypostasis in Maximus?

It might be interesting to note that Stephanus (the Alexandrian, Christian philosopher), in his commentary on Aristotle *De anima* book 3, says something on consciousness that might sound rather modern:

For the rational soul, according to them, does not have only five powers, intellect, thought, opinion, rational wish and choice; they add another sixth power to the rational soul, which they call 'attentive'. This attentive part, they say, stands over what happens in a human being and says 'I exercised intellect', 'I thought', 'I opined', 'I became spirited' 'I experienced desire'; and

[205] Törönen (2007), 56. [206] Ibid. 52. [207] Ibid. 99.

in general this attentive power of the rational soul ranges over all the powers, the rational, the non-rational and the vegetable.[208]

Of course, I do not claim that this commentary made any impression on Maximus. For one thing, it cannot be argued that he actually knew it. On the other hand he might have known the person who wrote these words, and he might have discussed similar things with him. Be that as it may, what strikes me is that a close reading of some Maximian texts might reveal traces of an idea of a hypostatic principle with a function similar to the 'attentive' part of Stephanus.

We turn first to some definitions Maximus gives in *Opuscula* 26 and 23. In the first he defines an individual; the second and third define the hypostasis:

1. An 'individual' is, according to the philosophers, a collection of properties, and this bundle cannot be contemplated in another; according to the Fathers, such are Peter or Paul, or someone else, each of whom is distinct from other men by virtue of their own, personal properties.
2. An 'hypostasis' is, according to the philosophers, an essence with properties; according to the Fathers, it is the particular man, who as person is distinct from other men.
3. For hypostasis is in all ways nature as well, like figure too is in all ways body. For hypostasis is not to be known without nature, and neither are figure or colour to be known without body. But nature is not in all ways hypostasis as well. For nature has the *logos* of being that is common, while hypostasis in addition has the *logos* of being that belongs to itself. The nature, then, has only the *logos* of the species, while the hypostasis is such that it in addition shows a someone.[209]

One might understand the first and second definition to indicate a movement away from the philosophical definition of *individual* to a more adequate conception of an individual being as a *hypostasis* (something distinguished by 'personal properties') according to the Fathers, but such an interpretation could be considered rather forced.

[208] Philoponus, *On Aristotle on the Soul*, trans. Charlton, 40. Charlton argues the commentary is by Stephanus.

[209] (1) *Th. pol.* 26, PG 91: 276a–b; (2) ibid.; (3) *Th. pol.* 23, PG 91: 264a–b.

One might wonder what is gained with the two definitions according to the Fathers. We could note that a comparison between definitions (1) and (2) shows that the terms hypostasis and person are synonyms. Further, an *individual*, according to Maximus, is someone who could be named, such as Peter. This someone is distinguished from other individuals by his own *personal properties*. What is characteristic of an individual, according to definition (1), is both that it is a nameable someone (τις) and that this someone is distinguished from others by his set of personal properties. Further, a *hypostasis* (2) is both a particular man and is distinct from other men as a person. It is quite obvious that 'individual' and 'hypostasis' denote the same thing, but not necessarily that they have exactly the same connotations.

The third text gives some additional information on the concept of hypostasis. Maximus defines hypostasis in relation to nature: hypostasis is in all ways nature. To help us understand this he suggests an analogy: figure is in all ways body, but body is not in all ways figure. The analogy runs like this: hypostasis: nature = figure: body. A figure does not exist in separation from a body, but is the shape in which a definite body occurs. Likewise, an hypostasis does not exist separated from a nature, but is always present as the hypostasis of *this* nature which it somehow delimits. With the presence of the figure or the hypostasis, the body or the nature is given a qualified presence. On the other hand, to be body is not the same as to be figure; likewise, to be nature is not identical with being a hypostasis. Body and nature may be understood and defined in the abstract, i.e. without reference to a definite shape and a certain hypostasis.

I believe Maximus' development of the understanding of hypostasis could be found in the traditional notion of *assumption* and in the principle of *selfhood*.[210] What I feel missing in Törönen's analysis of the Incarnation is the *principle* according to which Christ is *one*, i.e. one composite hypostasis the being of which is carried by the eternal hypostasis of the Logos Himself. If the charge of Nestorianism is to be avoided, something superior, some active ontological principle that is responsible for the total 'makeup', must be identified.

[210] In my thesis (Tollefsen 2000) I developed a more comprehensive argument, trying to make as much as possible of a supposed distinction between individual and hypostasis. I rather doubt now that this was a correct move.

In the Incarnation *God* became man in the whole of His being.[211] The *Logos* beyond being assumed being and possessed the concrete properties and movements characteristic of a human being. The one who initiates the assumptive process and is the centre of the whole dually constituted one being is the Logos Himself. He is the one who breathes and speaks, walks and uses His hands, is hungry and thirsty, in short possesses all human powers and faculties in a concrete mode, as His own (αὐτοῦ).[212]

In *Ambiguum* 5 there is talk of the Logos' ineffable assumpion (πρόσληψις) of human nature in the Incarnation.[213] In the *Disputation with Pyrrhus* Maximus says that Christ assumed (ἐδέξατο) what is proper to our nature.[214] That God became man in the whole of His being means that the Logos, when He assumed human form, appropriated *as His own* (ὡς ἰδίαν αὐτοῦ) the whole of what man is.[215] 'The hypostasis is in all ways nature.' Definition (3) above shows that the hypostasis has a *logos* of the kind of being or nature 'which belongs to itself' (τὸν τοῦ καθ' ἑαυτὸ εἶναι). The Logos appropriated a body and all the powers of the soul, i.e. intellect, rational will, powers of sensation, irascible, and appetitive faculties as the particular nature of His own self. As He appropriated them He became the subject or the *self* (αὐτός) including them all in His own consciousness.

Here we have to meet an objection from J. Meyendorff, who says that the hypostasis is not to be identified with something like ' "a center of consciousness", because Jesus, being fully man, possessed such a human center of conscious self-identification'.[216] Whether or not I agree with Meyendorff depends on what he means by a 'center of consciousness'. If Christ has two selves, both of them conscious of themselves, then it will be difficult to understand why He is one subject and not two. I cannot understand how we can avoid admitting that there are two persons in the God-man. On the other hand, He could well have 'a human center of self-identification' if what is meant is a human faculty by which the Logos can be humanly conscious of being human. This faculty would be the intellectual capacity of

[211] *Amb.* 5, CCSG 48: 19.13–14 (PG 91: 1048a).
[212] Ibid. 23.85–24.99 (PG 91: 1049c–d).
[213] Ibid. 21.36–41 (PG 91: 1048c). [214] *Pyrrh.*, PG 91: 297d.
[215] *Amb.* 5, CCSG 48: 23.85–9 (PG 91: 1049c).
[216] Meyendorff (1987), 18.

the soul, or maybe something like the attentive faculty of Stephanus. The 'self' (αὐτός) of this self-consciousness, however, must be the divine self of the Logos, and not the human self. The divine self has appropriated a particular human nature with all the human powers and all the categorical limitations which congrue on this nature as appropriated by the hypostasis, and united Himself hypostatically with it. He *is* man then, and His humanity is concretely His own. But there is only one self that would say about His divine as well as His human properties: I am these things, they belong to me.

Another objection, this time from Törönen, could be that self-consciousness is a natural, not a hypostatic property.[217] Once again one could quote Maximus: 'The hypostasis is in all ways nature.' One could say, of course, that self-consciousness is a feature of the species, but actually being self-conscious is something only particular hypostases can be. The natural possibility of being humanly self-conscious is assumed by the One who is the self of this concrete being.

What the hypostasis has assumed as itself is the particular nature which is the concrete instantiation of the specific nature. The nature as contracted to the hypostasis is the ontological condition of the one creature being different from the other within place and time. What is contracted is not, however, isolated from other beings, because ontologically the bond between beings is an ontic fact in the particular beings themselves. In the tension between this self-contraction and relatedness to others the creature moves on the basis of its internal structure of essence–potentiality–actuality, analysed above.

To get a closer view of the concreteness of hypostatic being we shall focus on an important distinction which is made by St Maximus. The being of the hypostasis is to be understood in the tension between the *logos* of nature (λόγος τῆς φύσεως) and the mode of existence (τρόπος ὑπάρξεως), which are often simply called *logos* and *tropos*. The terms mode and mode of existence denote a variety of different concretizations of nature. They denote the concretely manifested actualizations of the nature of the hypostasis, in its opinions, habits, acts, and its total mode of life. This always implies a hypostatic subject, although it is not identical in its ontological meaning with hypostatic being as such. This is why the concrete mode of human nature can exist in

[217] Törönen (2007), 56 ff.

Christ. It has to have a hypostasis, but not necessarily a hypostasis of its own kind.

A given hypostatic being cannot overstep the boundaries of its own natural distinction (ἡ φυσικὴ ἰδιότης), but its nature allows for considerable freedom with respect to the modes of hypostatic existence.[218] The possibilities which pertain to such modes are, of course, given along with the essential potentiality; they presuppose, however, that the creature is *not* determined by its essence to realize its potential in certain ways rather than others. A human being may, St Maximus says, 'give form to the mode of action':

> It is by being some thing (τι), not as being some one (τις), that each of us principally operates (ἐνεργεῖ) as man; but by being some one, e.g. Peter or Paul, he gives form to the mode of action (τὸν τῆς ἐνεργείας σχηματίζει τρόπον)—more or less intensively, determining it this way or that according to his gnomic will (κατὰ γνώμην). Hence in the *tropos* (ἐν μὲν τῷ τρόπῳ) the changeability of persons is known in their activity, in the *logos* (ἐν δὲ τῷ λόγῳ) in the inalterability of natural operation.[219]

The form man gives to the mode of action may be good or bad, and the mode of being of the hypostasis as a whole may accordingly be good or bad, i.e. virtuous or vicious. The natural will, always moving freely towards the good, may suffer modifications which determine the fate of the intelligent being by giving rise to a way of being which is out of harmony with the natural preordination of the *logos*. Instead of an essential movement in accordance with a being's *logos*, a movement in discord with its *logos* may arise and settle as a habit in the creature. This is exactly what happened in the fall of man. Man did not move as he should have according to nature, but began to move discordant with nature.[220] The breach with nature which this implies makes room for a kind of modified will which still moves within a range of possibilities, but the *situation* in which the subject chooses and the alternative options arise, is radically altered.

Before the fall, human existence was characterized by the unitary force of a free, natural will; after the fall this unitary force suffered

[218] Cf. *Amb.* 15, PG 91: 1217a; *Amb.* 42: 1341a. [219] *Th. pol.* 10, PG 91: 137a.

[220] The drama of these two possible movements is well illustrated by *Ambiguum* 41, cf. esp. PG 91: 1308c. Cf. also *Amb.* 10, PG 91: 1164c–d. The terms κατὰ φύσιν and παρὰ φύσιν occur quite frequently.

a split, because even though the natural will remained intact, it was nevertheless modified existentially when human beings in the post-lapsarian situation were confronted with subjective and objective conditions quite different from those of the pre-lapsarian situation. The original situation was characterized by a certain mode (*τρόπος*) of will which was in complete harmony with nature, and which gave the whole mode of existence of the hypostasis a God-willed direction. In this mode the hypostasis was in practical life centred upon the divine *logos* of its being, and was given the task of realizing the mediation between the cosmic divisions described in *Ambiguum* 41. The mode of fallen man, however, is—so to say—'eccentric', because it does not first and foremost orient itself by the divine *logos* in its way of being. The modified will in the insecure post-lapsarian situation is called *γνώμη*, 'gnomic will'.[221] One should not think, however, that the gnomic will is evil in itself. The point is that the fall came about through the exercise of gnomic will, and with this modified will man is exposed to the bewildering situations of decision-making, but by the gnomic will man may adjust himself also to a life in accordance with the *logos*.

Is this *logos/tropos* scheme characteristic of intelligent beings only, or does it concern other creatures as well? As far as I can see, this is a question which Maximus does not consider, but I believe it can be answered on Maximian principles. One should say that *only* human beings and angels have intellectual capacity to understand themselves in such a way as to be able to choose between alternative ways of being. The higher animals lack such a capacity, not to mention the rest of the animal kingdom and plants. But this is not all there is to be said, because it often happens that animals, plants, and natural phenomena generally do not behave 'normally' or 'naturally'. Of course, such a contention presupposes that one has a true knowledge of what is really *normal* or *natural*. Prey, hunting, killing, hostility, destruction, decay, transitoriness, illness, death, catastrophe—none of these things could reasonably be thought to belong to the original, divine scheme of the cosmic order, according to Maximus' understanding. This means that the behaviour of animals, plants, and natural phenomena is out of order, that is to say it does not follow modes in

221 Cf. Thunberg's analysis (1995), 213 ff.

accordance with the natural *logoi* of these things. Consequently, even if a wolf does not contemplate the possibility of achieving or not achieving its nature as a wolf, its mode of behaviour is determined by the modified conditions of the fallen cosmos. These conditions include a shortage of resources on which to feed, a struggle for survival, the fear of death, etc. According to this view, the unsatisfactory conditions of animal and plant life is a result of the human being's failure to achieve its natural existence in accordance with nature. According to the same line of thought, the whole 'natural' world is somehow perverted from its true nature and natural goal, in such a way that natural conditions turn into catastrophes which threaten to destroy living beings.

The topic of gnomic will sheds some further light on Maximus' understanding of the concreteness of being. Gnomic will is a 'a qualified act of willing, in relation to some real or assumed good'.[222] According to Maximus, the world has an original order in which a rational creature could easily move in accordance with its nature. The movement contrary to nature has perverted the relationship of the creature to the *logoi* of the world-order. In this perversion the creature is thrown into a situation it was not designed to cope with. The breakdown of the relationship to the natural framework makes the creature insecure and bewildered, because it has made the needs and passions arising out of its own self the norm of its behaviour.[223] Human needs and passions lie at the root of countless evils in social life and in man's relation to animals, plant life, and the so-called natural resources. In the post-lapsarian situation man is entangled in a net of occurrences. Now, my point here is to stress that if we are looking for a principle of *individuation* in Maximus, the gnomic will is certainly a strong candidate. The gnomic will is a dividing, yes, a disruptive factor in man's life.[224] It is no doubt that this will contributes considerably to the concreteness of man's being in the world.

These remarks seem to justify some important conclusions. The properties of man which result from his gnomic practices, could only

[222] *Pyrrh.*, PG 91: 308c. [223] Cf. *Or. Dom.*, CCSG 23: 343 ff.

[224] It is interesting to see that Meyendorff (1987), 22, considers sin to be a principle of individuation of the person.

be seen as individuating factors, but not as constituting his genuine *personal properties*. The moral character of a human being makes the total mode of existence of his hypostatic presence into something distinctive, but both the properties of vicious behaviour and the external characteristics concomitant with such behaviour are things man is called to combat and leave behind. Sinful practices and the socially and naturally established results of such practices are not destined to prevail. Through exercise of gnomic will, man is called to surmount what divides men, and to abolish differences established through the sinful execution of the gnomic will of fallen human beings.[225]

All of this does *not* mean, however, that individuality is something negative in itself. What I have argued above is that the sinful practices of the gnomic will are a principle of individuation, not that they are the only one. In the *Disputation with Pyrrhus* Maximus says that if human beings had moved according to nature, *one virtue* (μὶα ἀρετή) would have been visible in all.[226] This saying, one could claim, does not bear witness to any high evaluation of the *individuality* of personal being. On the other hand, there is no reason to believe that the one virtue does away with the genuine differentiation. Here we should remember one of the basic motifs of Maximus' philosophy, the unity-in-plurality that is destined to last forever.

We began this chapter with a focus on 'the mystery of Christ' and the incarnational motif. We shall return to this now. Maximus places great stress on God's intention to incarnate Himself. Through the *logoi*, the Logos incarnates Himself in created nature; and the Logos, as centre of the *logoi*, is the centre of the total cosmos, as we have seen. But this is not all. The Logos Himself is also the centre of each particular because each being is created by, and has its being from, the *logos* of its being *qua* particular. The *logoi*, we remember, represent the Logos to every created substance. In this way, the particular is both the final and the primary result of the creational process.

One of the most important lessons to be learned from this is that the particular being of each man has its *logos* from God, which *logos* is the centre of the person's very being. Every human person is

[225] Cf. the whole of *Amb*. 41, PG 91: 1304d–1316a, and of *Myst*. ch. 1, PG 91: 664d–668c.

[226] *Pyrrh*., PG 91: 309b.

therefore, in the depths of his existence, something willed by God. In this connection we can consider the question of whether God created me, this individual, or only the first members of the human species. Maximus does not, to my knowledge, address this question, but it should be rather easy to answer it on the basis of what we by now have seen of his doctrine. The answer is that God created not only the species, but distinct individuals as well. As a human person I was created by God. This does not mean that in the thought of Maximus secondary causes are completely abolished. But what is important to understand, is that in this way of thinking, nature, natural causes, and natural phenomena, are not only 'natural', as if the world functioned only on the basis of immanent principles. God is present in every natural process with His creative force. Instituting natural causality as such, He operates cooperatively to bring about what from eternity is conceived by Him in His *logoi*.[227] By *logos* and wisdom He *created* and *creates* (ποιήσας τε καὶ ποιῶν) everything at the proper time, Maximus says.[228]

Does this apply to human beings only? According to the principles of Maximus' metaphysics, we can surely answer that this is a universal truth: every particular being has its *logos* from God and is created by God.

A last point should be considered in this context. Does the concrete humanity of Christ have a *logos* of its own particular manhood? If the *logos* of the particular essence is the *logos* of the essence of a definite hypostasis, then the answer must be negative. There is no *logos* of a human hypostasis of Jesus. On the other hand, the humanity of Jesus has a principle of its own. Maximus says that the Logos in His Incarnation does not change anything that is human, but He constitutes it.[229] As far as I can understand, this means that while human hypostases have their principles in the *logoi*, the humanity of Christ has its principle in the Logos Himself: the Logos acts as *logos* of Jesus' concrete humanity. This has far-reaching implications. As the principle and the subject of Jesus' humanity, the incarnated Logos as centre of all the *logoi* is, in an extraordinary way, the microcosm which unites all beings. The redemptive actions of the Logos, consequently,

[227] Cf. the interesting section on providence in *Amb.* 10, PG 91: 1188c–1193c.

[228] *Amb.* 7, PG 91: 1080a.

[229] *Amb.* 5, CCSG 48: 21.36–41 (PG 91: 1048c).

have cosmic force. In Him the whole creation should be united in a general conversion (ἐπιστροφή) and achieve its glorification in a genuine unity-in-plurality according to the divine purpose.[230]

We started this chapter with the discussion of the image of the circle. In Maximus this image is not only an illustration of a metaphysical doctrine. It is an expression of a universal truth: Christ is at the centre of everything, both in the large 'organization' of the cosmos, in each and every created being, and in the letters, syllables, and sentences of Holy Scripture. In each and every being, in every detail of the cosmos, it is possible to detect Christ as centre, because His *logoi* are in each thing.

The importance of the doctrine of *logoi* should now be clearly seen. Some related points still remain to be considered, however. The first one has to do with the relation between the *logoi* and the divine 'activities' (ἐνέργειαι). Are they the same or do they differ? If they differ, what is the character of this difference, and how are they related to each other? These questions indicate the subject of Chapter 4. Therein we will also touch upon the important question of God's transcendence and immanence in relation to the world, a question which we have touched upon briefly in the present chapter, without, however, thematizing it directly. In Chapter 5, however, this subject shall get its due.

[230] Cf. *Amb.* 10, PG 91: 1188b–c; *Amb.* 41, 1308d ff.

4

The Divine Activity

The account of St Maximus' doctrine of divine *logoi* in the preceding chapter should not be seen as a complete one. The reason is that an adequate understanding of his doctrine presupposes that we see clearly both the distinctions between divine essence, activity (ἐνέργεια), *logoi*, and created beings, and the ways these elements are connected with one another. So far I have just taken for granted that there is a distinction between the divine *logoi* and the activities. In this chapter I shall discuss this distinction. The argument however, presupposes an understanding of what is meant by a divine activity, and my primary aim is to inquire into this topic. As the subject is developed, it will be possible to see in what the difference between *logoi* and activities consists. When this distinction is grasped, we become able to see exactly how activities and *logoi* are related.

Aristotle introduced the concept of *energeia*. According to him, the distinction between potentiality (δύναμις) and actuality or activity makes it possible to explain change and to argue against the Eleatic and Megaric denial of it.[1] The term *energeia* itself may have two senses. On the one hand, it may mean a motion (κίνησις). A motion is something incomplete (ἀτελῆς). It is the process towards some as yet unachieved goal, and it terminates when the work (τὸ ἔργον), which is established as 'outside of' the motion, is completed. On the other hand, an *energeia* in the proper sense is an action which includes the end, Aristotle says.[2] We may refer to the first type as 'incomplete activity', the second one could be termed 'complete activity'. We have

[1] *Metaph.* Θ, ch. 3. [2] *Metaph.* Θ, 3.1047^{a}30 ff., ch. 6.

above (Ch. 2) commented on the Neoplatonic doctrine of double activity, which could be seen as an adaptation of the original Aristotelian theory.

The expression 'a distinction between the essence and energies of God' sounds, I expect, a bit alarming. I suppose that everyone who has read the Introduction to the present work will expect me to argue a thesis that several scholars would call 'Palamitic'. In a way, this is what I am going to do. I shall not, however, take for granted that Maximus is a pre-Palamitic Palamist, even though he, in the end, may be found to develop a doctrine of divine essence and activities that is largely equivalent to the teaching of St Gregory Palamas. Since the Palamitic distinction plays an important role in the discussions of the present chapter, I shall first investigate its nature. This could seem to be a rather strange procedure, because, one could object, it is first important to see what the earlier Fathers themselves, Maximus included, have to say about divine activity before comparing their ideas with the doctrine of Palamas. However, for an adherent of the Palamitic tenet it could be highly tempting to describe the path from the Cappadocians via Maximus to Gregory Palamas as a teleological development towards a natural conclusion. It is possible for such an approach to view them all more or less as precursors to the great Hesychast doctor, to lose sight of the fact that earlier thinkers struggled with the problems of their own days, and not with problems belonging to fourteenth-century Byzantium. A chronological procedure, in addition to the temptation to write teleological, could prove to add to the obscurity of the whole subject. Readers who are unfamiliar with the distinction could get the impression that the author all the time keeps something to himself, because the final solution to the problematic and the most clear definition of the distinction will be given at the end of the discussion. However that may be, I choose the procedure of discussing Palmas first; to try to get as precise as possible an idea of the nature of the distinction as viewed by him (§ I), and then to ask whether a doctrine of a similar kind is present in the earlier Fathers (§ II). Then I move further to the distinction as understood by Maximus himself (§ III), before focusing on the relation between the activities and the *logoi* (§ IV). In the end this will furnish us with a coherent view of the thought of Maximus, with the 'mystery of Christ' at its centre.

I. GOD'S ESSENCE AND ACTIVITIES ACCORDING TO ST GREGORY PALAMAS

St Gregory Palamas (1296–1359) is an important figure in the development of eastern Christian thought. At the age of 20 he adopted the monastic life and moved to the Holy Mountain.[3] He developed his doctrine of a distinction between God's essence and activities in his defence of the monastic spirituality of hesychasm on Mount Athos. Barlaam, a monk from Calabria, attacked the hesychast method of prayer and the notion that what is experienced in prayer is the uncreated, divine light itself. The theological views of Palamas were endorsed in the Constantinople local councils of 1341, 1347, and 1351.

The point of departure for Gregory's doctrine was his wish to defend a certain practice and experience. If one denies that the light experienced by monks in prayer is the uncreated, divine light itself, because there does not exist such a light distinct from the very essence of God, then it follows (as one of several consequences) that there is no real deification of man, given that it is impossible for any creature to partake of the essence of God. Meyendorff says a real distinction between the divine essence and divine 'energy' is unavoidable in connection with the doctrine of deification, since deification implies participation of man in the uncreated life of God, and God's essence remains transcendent and totally unparticipable.[4]

Gregory Palamas argues that it is impossible to participate in the essence of God.[5] On this background he develops his doctrine of essence and activity in God. According to the quotation from Meyendorff a *real* distinction between the essence and the activity of God is unavoidable in view of the doctrine of deification, because deification implies a *participation* of created man in the uncreated life of God. If man is to be deified by participation in God, and if the essence of God is imparticipable, then man must be deified by participating some other 'aspect' of God than His essence. This other aspect is what is called the activity. But since the essence is imparticipable,

[3] For his biography and a detailed exposition of his doctrine, see Meyendorff (1974).

[4] Meyendorff (1987*a*), 186.

[5] These arguments are found in the *Capita 150*, 108–11.

the activity, as participable, must somehow be distinct from the essence. This distinction, according to Meyendorff, must be *real*. Now, what could it mean that the distinction is 'real'? 'Essence' means the immanent, self-identical being of God, while 'activity' means that God *does* something. Activity does not denote something other in God than essence, but is the same divine being as active. If, however, the distinction is supposed to be real, then essence and activity seem to denote different 'entities'. If that is the case, the question is what is the nature of the relation which these two entities have to one another?

The activity, Palamas says, is contemplated in God (ἐνθεωρεῖται τῷ Θεῷ), but God is not for that matter a composition (σύνθεσις).[6] Now, what is the meaning of this 'in'? One could think that this spatial metaphor should be taken in the sense that the activity is identical with the essence, but, as we have learnt already, this is not the case. I believe that what Gregory wants to emphasize is the fact that the activity is closely attached to the essence, and that it is not in any sense rather loosely related to the immanent being of God. When we say 'God' we denote not the trihypostatic essence separately, but the essence with its activity.[7] On the other hand, 'the essence with its activity' does not make God a synthetic being. The activity, Palamas says, is not separated from (μὴ χωριζομένην) the essence, because it is from it (ἐξ ἐκείνης οὖσαν). The correct thing to say is that the activity *differs from* the essence (διαφέρει τῆς οὐσίας ἡ ἐνέργεια).[8] When Meyendorff says that the distinction is *real*, this has to be balanced against the sayings of Palamas that the activity *differs from* the essence, but is *not separated from* it. I wonder if the word *real* in the term real distinction could not give unhappy connotations.

The essence and the activity are in a certain way different from each other. Now, perhaps this distinction may be understood along the lines of Aristotelian logic? Gregory actually asks whether the activity could be an *accident* (συμβεβηκός), maybe even a so-called 'inseparable accident'.[9] However, the answer is negative. An accident in the ordinary sense comes into being and passes away again. The activity of God does not have this character. That the activity should be an

[6] *Capita 150*, 128. [7] *Ibid.* 126, 145, cf. 109.
[8] *Ibid.* 126, cf. *Triads* 3.2.13. [9] *Ibid.* 127 and 135.

inseparable accident is also denied, even though without any reasons given, by Palamas. According to Porphyry's *Isagoge*, an inseparable accident does not belong to one species only, as the capacity to laugh belongs to man and to man only. It may belong to different species, just as black belongs to the Ethiopian, but also to the crow, etc.[10] However, Gregory does not drop the idea that the activity has some kind of accidental character. He points to the fact of it being called a *kind of accident* (συμβεβηκός πως) by some theologians. Given that the activity belongs to God in a natural or proper sense, could not this kind of accident be understood as a *property* of God? We shall examine this a bit closer.

In the *Isagoge* Porphyry distinguishes between four kinds of properties.[11] The last kind of property is property in the strict sense, according to Porphyry, because it is convertible, as it occurs in the whole species, in it only, and always. In philosophical terminology, from Aristotle to Late Antiquity, three terms are used for property, viz. τὸ ἴδιον, ἰδίωμα, and ἰδιότης.[12] What is important in our context is that property in Aristotle and Porphyry does not indicate the definition (ὁ ὅρος) of the 'whatness' (τί ἐστι) or the 'essence' (τὸ τί ἦν εἶναι) of a being (οὐσία). The capacities to learn grammar and to laugh may manifest themselves in certain *activities*, such as actually laughing and learning grammar. These capacities and activities belong naturally to

[10] Porphyry: *Isagoge*, CAG 4.1: 22.5 ff.

[11] Ibid., CAG 4.1: 12.13 ff.

[12] The three terms are often used in the same sense. It seems that ἰδιώμα and ἰδιότης are used more frequently in Christian thought in Late Antiquity than τὸ ἴδιον, while the latter term is more frequently used by Aristotle and Porphyry. When it comes to Aristotle, Bonitz' *Index aristotelicus* lists only one instance of ἰδιώμα. Unfortunately this is from *De plantis* (821^{b}22). The text shows that the term has the same sense as τὸ ἴδιον in Porphyry's second sense of the term. But even if *De plantis* should be a work by Aristotle originally, the present text is a translation of a medieval, Latin edition. In the original the author could well have used τὸ ἴδιον instead of ἰδιώμα. The index of the CAG edition of Porphyry's *In cat.* has only one instance of ἰδιώμα (55.19). It seems to be used in the fourth sense of property listed by Porphyry. Bonitz has two instances of ἰδιότης, one in the *De plantis* (822^{a}4) and another in the *De mirabilibus auscultationibus* (836^{b}23). The *De mirabilibus auscultationibus* is not by Aristotle, but it may have originated in the Peripatetic school. The index of CAG has four instances of ἰδιότης in Porphyry's *Isagoge,* and *In cat.*, 7.22 ff. shows that it may be used in all four senses listed by Porphyry. In 20.7 and 22.11 it means 'property' of other types of predicates. In 129.9 l. it is used about the sum of characteristics which makes one individual different from another.

man always, but they are distinct from the essence of man, because the human being is defined without taking these properties into consideration.

If the divine activity can be taken as a property in the sense outlined above, it has, however, certain peculiar characteristics. First, it is a one-many; and, second, it is an object of participation. Its character as one-many is related to its participability, which we shall return to below. First I shall make some comments on the second characteristic.

For Gregory Palamas, the terminology of participation obviously has strong material connotations. The participated substance is somehow divided in parts which are received by the participants. Palamas, of course, does not consider the divine activity as some kind of material substance, but even so, it is difficult not to get the impression that some thinkers who operate with a concept of participation somehow allow for an idea of a quasi-material 'something', at least as an imperfect image of a metaphysical truth. To illustrate this, it should be enough to point to the so-called 'doctrine of emanation' in Neoplatonism, which clearly is an image, but which has been mistaken for actual doctrine by many scholars.[13]

From the works of Gregory it is easy to see that the primary sense of *energeia* is *activity*. The *energeia*, he says, quoting St John of Damascus, is 'the essential motion of nature' (ἡ οὐσιώδης τῆς φύσεως κίνησις).[14] Maybe this is a much better approach to the topic of divine activity than the questions as to what kind of predicate it might belong (i.e. accident, property etc.). The capacity for activity belongs to the nature from which it proceeds. As examples of divine activity he mentions that God *foreknows* and *provides* for inferior beings, that He *creates*, *preserves*, *rules*, and *transforms* them (προγινώσκει, προνοεῖται, δημιουργεῖ, συντηρεῖ, δεσπόζει, μετασκευάζει).[15] The activities, Gregory says, this time quoting Dionysius, are certain powers (δυνάμεις) which are deifying, causing being, giving life, and granting wisdom (ἐκθεωτικὰς ἢ οὐσιοποιοὺς ἢ ζωογόνους ἢ σοφοδώρους).[16] All this emphasizes the dynamic character of these activities.

[13] Cf. Gersh (1978), 17 ff. [14] *Capita 150*, 143.

[15] Ibid. 137, cf. 136 for the activities of human beings, which, according to Palamas, are distinct from their essence.

[16] *Triads* 3.2.11; cf. Dionysius, *DN* 2.7, PG 3: 645a.

When the activity is spoken of 'objectively' as a divine work (ἔργον) or being,[17] I think we are not to imagine a something existing between the divine essence and creatures. The terms work and being denote, I suppose, the *reality* of God's activity as a powerful presence. As we shall see below, this resounds with Maximian terminology. Palamas denies that the activities are many gods or spirits, i.e. that they could be hypostasized. Rather they are the processions, manifestations, and *natural* activities of the Spirit.[18] This resounds with Dionysian terminology. Gregory several times stresses the *uncreated* character of the activity. God's activity is not something that begins and ends, but is a *permanent* expression of the divine being itself. This, however, is difficult to understand, and I shall try to expose what I feel could be a problem.

If we consider the workings of the activity mentioned above, it becomes clear that God as creating, providing, deifying, etc., is God *in relation to* His creatures. That there is a distinction between the essence and the activity in this respect seems clear. God, as He is in Himself, in His own essence, must be distinguished from God as He relates to something other than Himself. To say that God is by essence a creator would be to determine the being of God, not as something in itself, but in relation to what is not Himself. This line of thought should be quite in order, but the idea of the activity as uncreated and permanent seems to create a problem. If creatures have a beginning, and God's relation to creatures is established at the moment when they begin to emerge, is it not from then on *only* that it becomes meaningful to speak of a divine activity *ad extra*? Could the activity be understood in some way as a *permanent property* of God? Or is it not rather of an accidental character? Of course, Gregory himself and the Palamists would deny that. The activity belongs eternally to God as the natural property of His being. This means that the activities somehow are *proper* to God's essence even 'before' God relates Himself to anything 'other' through them. Only at the moment which the divine will eternally willed to be the beginning of creatures, is the activity manifested *ad extra* as creative, providing, deifying, etc. As the eternal property of God, the activity is potent with all kinds

[17] Palamas, *Triads* 3.2.6 and 7; cf. Maximus, *Cap. gnost.* 1.48, PG 90: 1100c.
[18] *Capita 150*, 71.

of possible effects, which God could have. To exist in such a way, according to this view, is natural for God. But only at the 'moment' decided by the divine will did the potency become an activity in relation to something other.

Perhaps we could seek some help from the Plotinian terminology of double activity. In the eternal generation of the Son and the procession of the Spirit the three divine hypostases are intimately *related* to one another. In this relatedness a certain immanent or internal activity is manifested. The character of this activity could be said to be the mutual love which the hypostases express towards one another. On the one hand, this internal activity, as the natural expression of the one essence through three loving persons, should be seen as intransitive. It does not act *ad extra*, because there is as yet no *extra*. On the other hand, when the divine will is joined to this natural effluence of the internally manifested divine love, the activity becomes out of the essence. This way of interpreting the activity does justice to the Palamitic idea that for a creature to receive its share in the activity—however problematic the language of participation may be—is to receive a share in the divine life. I think that viewed along these lines the Palamitic distinction is philosophically understandable, even though the 'mystery' of participation still remains to be solved.

That creatures participate in the divine activities must mean that God works in them, and this activity in creatures takes place on two levels, viz. on the cosmological and the soteriological level. Cosmologically God is active in beings by giving them being, and soteriologically, He is active by granting deification. In *Capita 150*, after he has stressed God's transcendence, Palamas says that God is the nature of all beings since all participate in Him and receive their constitution by this participation. They participate in His activity, not in His nature: 'Thus is He the very being of beings and the form in the forms as the primal form and the wisdom of the wise and generally all things of all things.'[19]

Some lines below this he clearly distinguishes between the natural participation which is common to all creatures, and the participation which is granted to creatures that are capable of drawing near to God

[19] *Capita 150*, 78, trans. Sinkewicz, in Palamas *The One Hundred and Fifty Chapters*.

by the exercise of will. By being active in creatures on the cosmological level God becomes the being of beings and their formative principle. On the soteriological level He becomes the wisdom of the wise, etc. It is important to note, however, that deification not only consists in having virtue and wisdom, but that virtue and wisdom manifest the participated activity.[20] One could wonder what it means that God becomes the being of beings and the wisdom of the wise. Does it mean that the being of beings is not a created being, but the divine activity itself? I think that the answer is positive, but this does not mean that the divine and the created spheres are confused. According to Palamas there is a created otherness. On the condition that it is the same logic which regulates the relations on the cosmological and the soteriological level, this is clearly shown when he speaks about deification. Even though the creature receives a share in divine attributes, it continues in its creaturely status when deified.[21]

As something to be participated in, the activity shows its character as a one-many. The divine essence is one and altogether indivisible (*μία γὰρ ἐκείνη καὶ παντάπασιν ἀδιαίρετος*). Since the 'illuminations', i.e. the activities, are a plurality, it follows, according to Gregory's argument, that they are distinguished from the essence.[22] This seems to be a basic idea, viz. that the difference between essence and activity is demonstrated by the plural nature of the latter. However, the activity is not just a plurality, it is characterized by being 'divided indivisibly' (*μεριζομένη ἀμερίστως*), Palamas says.[23] Once again we have an expression resounding with a Maximian idea (as we shall see in Ch. 5 § IV). Palamas tries to explain this 'divided indivisibly' by the use of the analogy of the sun's ray, which is one, but gives warmth, light, life, increase, etc. By analogy we can understand this to mean that the activity is *one* by its nature, but *many* in its effects in relation to creatures. The one activity is God's simple power which in relation to creatures may be participated in a manifold of aspects. The activity's being *divided in an indivisible way* could mean exactly this, that it is present according to its indivisible nature establishing the participating creatures in different relations to itself and thus being 'divided'.

[20] *Triads* 3.1.31. [21] Ibid. [22] *Capita 150*, 65.
[23] Ibid. 68; cf. 69, 74, 81.

When Gregory says that the activity is bestowed proportionally upon the participants according to the capacity of the receivers (κατὰ τὴν ἐπιτηδειότητα τῶν ὑποδεχομένων),[24] it could seem that the diversification of the activity depends upon something *other* than God's creative scheme. But then the receivers would, to a certain degree, have an independent role in the constitution of the cosmos, and this does not seem probable. The activity's plurality is not dependent upon the different ways in which it is received by different beings, but it becomes effective of different results and is received in different ways in accordance with the divine paradigms for different beings and existential modes of these beings. The sun's ray, being one, effects several things. In an analogous manner the activity, being undivided, is divided *as* effecting several things in relation to creatures. In the context of this problem there seems to lack a more elaborate and nuanced application of principles in Gregory. What are the reasons why the activity is participated proportionally? Palamas explicitly answers this question by pointing to the divine will as the principle of distribution,[25] but he does not seem to have developed a doctrine of *logoi* as acts of will, functioning as distributive principles, in the way St Maximus did. Rather, Palamas identifies the activities and the *logoi*, which immediately seems rather unfortunate.[26] However, when he stresses the distributive role of the divine will, the problem should in principle be solved.

Before we leave Gregory Palamas, some words should be added about his interpretation of his predecessors. Even though he argues philosophically for the distinction, a thoroughgoing mark of his exposition is the documentation from the Fathers and the arguments from authority. Are the interpretations he gives of his predecessors sound? What is striking is that Palamas does not seem to press his sources or to over-interpret them. For instance, when he appeals to St Basil the Great, Palamas feels quite satisfied with the exact wording of his authority he just cites Basil and draws simple lessons from his words.[27] In general, the ways the predecessors express themselves fit, at least apparently, into the picture as Palamas sees it.

[24] Ibid. 69. [25] Ibid. 91. [26] Cf. *Capita 150*, 87; *Triads* 3.3.10.

[27] Cf. examples of his interpretation of St Basil in *Capita 150*, 68 and 72, of St Maximus in *Capita 150*, 111 and *Triads* 3.2.7. Palamas has many references to Basil and Maximus, and of course to several other Fathers.

However, even if this is the case, it is always possible that Gregory Palamas tries to argue a thesis that his predecessors did not intend to argue. Even if the terminology does not differ much, the matter of fact might differ. The Cappadocian Fathers, Dionysius the Areopagite, and Maximus did not have the 'Palamitic' problem in view. Their theology and philosophy were not developed to secure the genuine character of the mystical experience against adversaries. Of course, this does not preclude in advance that the distinction between essence and activity is not present in any of the earlier Fathers, even though they must all be seen as addressing problems of their own day. Before we move to our main subject of investigation, we shall take a look at some important thinkers before Maximus just to consider what could be the meaning of essence and activity in them.

II. THE DOCTRINE OF ESSENCE AND ACTIVITY PRIOR TO ST MAXIMUS

St Basil the Great wrote a famous letter to Amphilochius to which some of those who defend the doctrine of a distinction between the essence and energies of God appeal.[28] There is, however, reason to ask whether such an appeal is justified. First of all we must see what Basil actually says in the letter. There is no doubt that he distinguishes between the divine essence and activities. He says:

> For they themselves have confessed that essence is one thing and each of what was enumerated was another. 'Nay, the activities are varied and the essence is simple.' But we say that from His activities we recognize our God, but His essence itself we do not profess to approach. For His activities descend to us, but His essence remains inaccessible.

I shall return below to the important question of who 'they themselves' are. The 'what was enumerated' refers to a list of attributes of a dynamic kind earlier in the letter. This list has a polemical sting, because Basil's opponent confesses that God is *simple* and that all knowable attributes are reduced to the simple divine essence.

[28] *Letter* 234 in the Loeb edition of 1986 (3: 371 ff.). Appeal is made, e.g. by Lossky (1973), 71–2 and Habra (1957–8), 297.

Addressing this challenge, Basil asks, given the idea of simplicity, whether divine fearfulness and benevolence are the same, justice and creative power, foreknowledge and requiting, magnificence and providence. Obviously there are tensions here, for instance between the divine fearfulness and benevolence, and it seems justifiable to doubt whether these can be reduced to one and the same, i.e. be just one simple thing. The point of all this seems to be that, according to Basil, the simple divine essence is not identified with the activities. The activities make it possible to recognize *that* God exists, but not to know *what* He is.[29] If, on the other hand, we confess that we *know* God (τὸν Θεὸν ἐπίστασθαι), this would be in the sense that we *understand* the scriptural testimony of God's deeds. That this is what is intended, seems to be confirmed by the last part of the letter, where Basil points to the divine activities on the historical scene, as witnessed by the Scriptures.

The point in this is that God is unknowable in Himself, but from His activities we may (i) recognize His existence and (ii) know Him as He who is active in relation to His creatures. In this letter *energeia* means 'activity', and no conclusions can be drawn as to whether Basil considers this to be *really* distinct from the essence, and no Palamitic problem is in the horizon.

Who are Basil's adversaries? They are Eunomians, the followers of Eunomius, the neo-Arian theologian; the letter should therefore be interpreted in light of the controversy between the Cappadocians and the Eunomians.

According to St Gregory of Nyssa in his *Contra Eunomium*, Eunomius teaches a system of three essences, the first one is the supreme and absolute essence, the second one exists by reason of the first and the third one is inferior to them both; that is, inferior 'to the one, as to its cause, to the other, as to the activity which produced it'.[30] This means that the first essence is the original cause of everything, but it is the second essence which produces the third by the activity it has received from the first.

[29] ἐγὼ δέ, ὅτι μὲν ἔστιν οἶδα, τί δὲ ἡ οὐσία ὑπὲρ διάνοιαν τίθεμαι. Deferrari (Loeb) understands ἐκ τῶν ἐνεργειῶν γνωρίζειν τὸν Θεόν as 'to know God from the energies', but Edgars Narkevics (Riga) has suggested to me that it should be translated as 'to recognize God from the energies'. I agree with Narkevics.

[30] *Contra Eunomium*, ed. Jaeger (1960), 1: 72. 5–6.

According to the description of Eunomianism given by Gregory, each essence is followed by activities, and 'its activities are bounded by its works, and its works commensurate with the acticvities'.[31] This doctrine justifies the further Eunomian doctrine of the knowableness of God.

Eunomius says that the activities follow the beings and are relatively greater and less. The difference amounts to that existing between their works. According to Eunomius, it is not correct to say that the same activities produced the angels and the stars, the heavens and men. In proportion as some works are superior and more honourable than others, so does one activity transcend another in dignity.[32] It should be easy to understand the idea that the work is commensurate with the activity which produced it, and that different activities are needed to produce different kinds of work. One could say that a product of a craft is commensurate with the skill of the one who produced it, and that the work itself is neither more nor less than this skill. This principle, however, is not allowed by Gregory, and I believe he is correct in denying it. He says that all the resources of a smith are not set in motion to make a gimlet.[33] Even if the limit of the activity is somehow found in the work, this does not mean that the nature of the activity or the nature of the agent which exerts it are fully revealed in this product.

Even though Gregory argues against Eunomius, both Gregory and his brother accept the concept of activity. In the *Ad Eustathium de sancta Trinitate* Gregory defends the divinity of the Holy Spirit. His opponents asserted that 'God' is indicative of the nature, and should therefore not be applied to the Spirit because He does not have His nature in community with the Father and the Son. To this Gregory answers that the nature is exalted above human understanding, so that we have to argue from some observable evidence about a being which evades our knowledge. The nature of God, consequently, can only be investigated by its activities. If the activities of the Father and the Son and the Holy Spirit differ from one another, then the natures, which operate, are also different. But all divine attributes, he argues, even if they have different connotations, denote the same thing. If we ascribe several such attributes to the Holy Spirit, it naturally follows

[31] Ibid. [32] Ibid. 1: 72.10 ff. [33] Ibid. 1: 150.11 ff.

that we should also ascribe to Him the appellation of 'God', and therefore the three persons have the same essence. The essence itself, however, remains unknowable, but we may know that He is a judge, that He is beneficent, good, just, and all else of the same kind. These are diverse activities, and we do not know His nature from these.[34]

As I have said above, the defenders of the Palamitic distinction between essence and energies appeal to St Basil's letter to Amphilochius. The texts we have considered from St Gregory would likewise count as evidence for this distinction. Is such a position tenable? From what we have seen so far, the concept of divine activity is, in these two Cappadocians, connected with the possibility of attaining *knowledge* of God. The distinction between essence and activity must be understood in relation to the contemporary polemics with the Eunomians.

However, as I said in the preceding section (§ 1), Palamas' interpretation of the Fathers does not seem forced. He has no problem in finding and adopting a whole arsenal of textual confirmations of his doctrine. I believe that several of his applications are sound, for instance those which are taken from Dionysius and St Maximus. When it comes to the Cappadocians, however, the situation might seem to be a somewhat different, since their essence/activity distinction may seem to be given a full explanation without mixing in my notion of Palamism. From a scholarly point of view every thinker must be interpreted within the limits of his own historical setting. But if the Cappadocians actually accepted a distinction between the essence and activities of God in the context of a particular controversy, the theological gain of this doctrine could be of much wider significance than the Cappadocians themselves were aware of in that special situation. Even if the Cappadocians themselves did not have a distinction of the Palamitic kind in mind, what they actually say is consistent with the thought of St Gregory Palamas.

This, however, is not necessarily the end of the matter, for if the distinction is applied to explain cases with ontological implications, the case for a Palamitic interpretation could be strengthened. I can imagine at least two cases, which could have such implications, viz. in

[34] *Ad Eustathium*, in *Gregorii Nysseni Opera*, ed. W. Jaeger, vol. 3, part I (*Opera dogmatica minora*), ed. F. Mueller (Leiden 1958), 1–16.

connection with cosmology and soteriology. To make myself clear, if the distinction is used to explain *participation* in God in creation and deification, then we come close to Palamas' concern. A condition for this is of course that the Cappadocians realized what Meyendorff saw as a basic truth: that it is impossible to participate in the essence of God.

Against the contention that participation entails a *real* distinction between the essence and the activities of God, one could object that the need to understand the distinction as 'real' or not depends on what one understands by 'participation'. If participation is interpreted the way David Balás (1966) did, then a real distinction is an obvious fact, because what is participated in seems to be a created perfection. A distinction of the Palamitic kind is, however, not relevant. As far as I understand Balás, to participate means to have a created perfection, which has some kind of likeness to a divine attribute. But, in my view, this is not exactly the way participation is understood in the Greek Fathers. What I would like to investigate now is whether there are texts in which the essence/activity distinction is used in cases which have ontological implications.

The first passage is from chapter 9 of St Basil's *De Spiritu Sancto.*[35] The first impression one gets from reading through this chapter is that when it comes to essence and activity Basil's thought moves along the same lines as in his letter to Amphilochius. This is well illustrated from the following propositions about the Spirit: 'By nature He is unapproachable, He is apprehended by reason of goodness...He is simple in essence, manifold in powers.' So far there the need is to distinguish between the nature or the essence of the Spirit, and His economical activities. The question now is how these 'economical activities' work, and we are immediately presented with the two related topics of participation and deification. 'He is impassively divided, shared while remaining whole', Basil says. The power, the activity, or the grace is 'something' participated in by rational creatures. The renewal of the image of God in man is connected with the presence of the Spirit. When the image is renewed, man, with purified eyes, will see the beauty of the archetype in the Spirit. The light of the Spirit will fall on human souls, making them Spirit-bearing

[35] PG 32: 108a–109d.

souls (πνευματοφόροι ψυχαί), with the result that the souls themselves become spiritual (πνευματικαί). The dividing and sharing out of the Spirit is His sharing out of the many powers by which He acts in creatures. This activity seems to become established as a presence in the souls, so that they are made Spirit-bearing. The final result of this regenerative process is that man is to achieve the *likeness* with God (ἡ πρὸς Θεὸν ὁμοίωσις), yes, is to become God (Θεὸν γενέσθαι).

We shall turn next to St Gregory of Nyssa. In the *De vita Moysis* he says that God is participated in by all, without being lessened by their participation.[36] According to Balás, it is seldom that Gregory *explicitly* employs the terminology of participation when he speaks about the dependence of the created world on God.[37] Usually the dependence is expressed in terms of God *permeating*, *being present in* and *enveloping* the cosmos. These ideas are prominent in Gregory. In the *De anima et resurrectione* he writes about a divine Power which permeates all things, fits the parts together with the whole and fulfils the whole in the parts.[38] This power maintains everything, remains in itself and revolves around itself. Gregory also says that the ineffable wisdom of God which appears in the cosmos, gives us to understand that the divine nature and power is in every existing thing.[39] Because of this presence all things remain in being. In *Oratio catechetica* he says that the divine is present in everything. It penetrates, embraces and is seated in it.[40] All things depend on 'He Who is', and nothing exists which does not have its being in God. All things are in Him, and He is in all things.[41]

Now, in what way is this relevant for the distinction between God's essence and activities? First we should note that some of this sounds rather strange as the utterauce of an orthodox Church Father. It somehow has a pantheistic ring to it. One could of course argue that the more pantheistic expressions are to be taken figuratively—hardly a fruitful line of interpretation. But in fact this pantheistic language is not at all what is seems to be: Gregory could simply be thinking

[36] *De vita Moysis* 2.25. Greek text in Grégoire de Nysse, *La vie de Moïse*, ed. J. Daniélov Sources Chrétiennes (Paris 1968).

[37] Balás (1966), 115–16.

[38] *De an. et res.*, PG. 46: 28a.

[39] Ibid., PG. 46: 44a–b.

[40] *Or. cat.* ch. 25, *PG*. 45: 65c–68a.

[41] Cf. *Or. cat.* ch. 4 and 32, PG 45: 20a–c, 77d ff.; *De vita Moysis* 2. 174, 177, and 179.

in terms of essence and activity. To make myself clear: all this talk of permeation, presence, and envelopment could just mean that God is *active* in all regions of the world, creating, maintaining, and directing all natural phenomena. Even the terminology of participation, when it occurs, could be taken in the sense that to participate in God means that God is present as actively working or present in His activities. When it is said in the *De anima et resurreccione* that the divine essence is totally different from creatures, but that it is still present in created beings, Gregory may well mean that we have to distinguish between God's essence and His activities, between what He is and what He does.[42]

Gregory's language very much invites an orthodox interpreter to discover a distinction of the Palamitic kind. Gregory often expresses the dialectic between transcendence and immanence. God created the world, he says, so that man through his mind (νοῦς), yes, so that all things, may participate in the beautiful and the good.[43] At the same time God transcends even Beauty and Goodness.[44] God is by nature life-giving, but He transcends at the same time all characteristics. It is even a characteristic (γνώρισμα) of the divine nature to transcend all characteristics (παντὸς ὑπερκεῖσθαι γνωρίσματος). God, therefore, is not something to be known.[45] According to Gregory, God is True Being (τὸ ὄντως ὄν), and all things exist by participation in divine being.

Adherents of the Palamitic distinction would often take this dialectic between transcendence and immanence as an indication of a distinction between the essence and the energy (activity) of God. Why should this not be possible? Because, says Balás, 'the repeated insistence of Gregory on the simplicity of God would not allow this'.[46] Balás' well-known book, with the promising title *ΜΕΤΟΥΣΙΑ ΘΕΟΥ, Man's Participation in God's Perfections according to Saint Gregory of Nyssa*, contains much valuable material and several thought-provoking analyses. Balás interprets Gregory's doctrine in the context of the latter's anti-Eunomian polemics. He concentrates

[42] *De an. et res.*, PG. 46: 44a.

[43] *De hom. op.* ch. 12, PG. 44: 161c; *Or. cat.* ch. 6, PG. 45: 25c.

[44] *De hom. op.* ch. 16, PG. 44: 184a; *De an. et res.* PG. 46: 93a; *De virg.* ch. 10, PG. 46: 361a–b.

[45] *De vita Moysis* 2.234, cf. 2.176.

[46] Balás (1966), 128.

his study of participation on the three divine perfections of Goodness, Life, and Being.[47]

Balás denies any Palamism in St Gregory of Nyssa. He admits (i) that the divine essence transcends its attributes, and (ii) that beings participate in the perfections. (The terms 'attribute' and 'perfection' seem to be synonymous.) Palamists would *agree*, however, both on the simplicity of God, and that the multiplicity is found only in the created effects and in our conceptual knowledge starting from these effects, and not in the divinity itself. So, why is a Palamitic interpretation not allowed? Could it be that there is some difference of opinion upon exactly *what* is participated? Balás is not clear on this point, so what I am constructing is no more than a hypothesis with some probability.

Balás says that the principle of participation 'refers primarily to the composition between the subject and its own (participated) perfection'.[48] What ontological status does 'its own (participated) perfection' possess? Balás says that the term τὸ μετεχόμενον may have two meanings, 'it may refer to the transcendent perfection' and 'it may refer also to these perfections as possessed by us, to our goodness, virtue, life, etc.'[49] In his Conclusion Balás says that God is present in each participant 'as the efficient and especially exemplary cause of the shared perfection'.[50] Now, taken together, this probably means that participating, in the sense of being present in a creature as a 'part' of it, is a quality ('its own [participated] perfection') *created* by divine efficient causality. To participate then, means to possess certain created perfections, which shall be developed in a life in accordance with the divine intention. In this way God is exemplary cause too, viz. in the sense that the creature will constantly increase its 'share' of the perfection in keeping God before its eyes.

If this interpretation is correct, we might reasonably ask if Balás is not sceptical from the outset about the Palamitic understanding of Gregory of Nyssa. If what I have described actually is Balás's view, then I disagree with his interpretation, and remain in my so-called Palamitic position.

Balás however, in a note, refers to a text in the *Contra Eunomium* in which the distinction between essence and activities is 'explicitly

[47] Ibid., chs. 2–4. [48] Ibid. 130. [49] Ibid. 129. [50] Ibid. 163.

rejected by Gregory'.[51] I agree that the text at first seems to reject the distinction, but I am not sure that a closer examination will confirm Balás' view. Gregory is critical of Eunomius' doctrine that the activities 'follow' (ἕπονται) the essences. The question is what kind of relation between essence and activities Gregory thinks is implied by the verb 'follow'. It seems that the relation for instance between fire and its heat is intended to illustrate the kind of relation against which Gregory is arguing. But the problem with this illustration is that in another text he employs it to exemplify his own understanding of the relation between essence and activities.[52] Gregory seems to deny two things, (i) that the activity is somehow 'outside of' the being it belongs to, and (ii) that it makes this being to be 'something manifold and composite' (ποικίλον τι χρῆμα καὶ σύντετον). Gregory himself has a doctrine of the activities as existing 'around Him' (περὶ αὐτό) so that the force of the 'follow' could not be to deny that they are 'around' the essence of God.[53] Rather he wants to deny that the activities are established as a kind of external reality in themselves. A Palamist, even if he speaks about the 'energy' as 'outside of the essence', does not mean that it is established as some kind of quasi-hypostasis 'between' the essence and the things on which it operates. The activity does not 'follow' the essence in this external fashion. A Palamist would also deny that the distinction makes God into a 'manifold and composite' being, as I asserted above. As far as I can see, even though the text could at first sight be taken to imply what Balás thinks, it is not incompatible with an idea of a distinction between essence and activities.

My interpretation can be further confirmed by a study of Gregory's sixth sermon on the Beatitudes.[54] The text is from Matthew 5:8: 'Blessed are the pure in heart; for they shall see God.' Should any subject bring up the need to distinguish between essence and activities it would have to be this one, for how is it possible to *see* God? In the Scriptures, Gregory remarks, both John and Paul deny that a vision of God is possible.[55] But, Gregory says, one may 'see'

[51] Ibid. 128 n. 48. *Contra Eunomium*, ed. Jaeger (1960), 1: 87.3 ff.

[52] Cf. *Ad Eustathium de sancta Trinitate*, Jaeger and Mueller eds. (1958), 3.1: 11.9–10.

[53] Balás (1966), 110.

[54] *De beat.*, PG 44: 1264b ff.

[55] Cf. John 1:18 and 1 Tim. 6:16.

Him who made all things in wisdom by an inference from the wisdom that appears in the cosmos. God is then 'seen' with spiritual eyes. In human works of art the mind can perceive the maker of the product because he has left on his work the stamp of his art. This does not mean, however, that what we see is the *nature* of the artist, but only his *artistic skill*, which he has impressed on his work. It will be in the same way if we look to the order of creation. We form in our minds a concept, not about the essence, but about the Wisdom (οὐ τῆς οὐσίας, ἀλλὰ τῆς σοφίας) of Him who has made all things wisely.[56] We may also apprehend His Goodness, but again not His essence (τῆς ἀγαθότητος οὐ τῆς οὐσίας).[57] Gregory says further: 'For He is invisible by nature, but becomes visible in His activities, for He may be contemplated in the things that are around (περί) Him.'[58] So far it could seem that this has to do with 'seeing' God, that is with *attaining knowledge* of Him. But 'to see' (ἰδεῖν), Gregory says, may in the Scriptures mean the same as 'to have' (σχεῖν).[59] To see God in the purity of the heart then, means to participate in Him. Beatitude does not consist in knowing something about God, but to have God present in oneself.[60]

Based on Gregory's argument it is impossible to *see*, that is to *have* the divine essence. But it is possible to *see*, that is to *have* the divine activities in oneself. The way Gregory formulates the relation between the essence and the activities here obviously has to do with a distinction which makes it possible for created beings to participate in God without having to participate in His essence.

Before turning to St Maximus the Confessor, we shall see how Dionysius the Areopagite has an explicit idea of God's manifestations *ad extra*. It is well known that Dionysius considers the Divinity in itself as hidden in its unattainable and unknowable transcendent majesty. God is called 'the unparticipated Cause' (ὁ ἀμέθεκτος αἴτιος).[61] On the other hand, the Areopagite strongly emphasizes that God is accessible and participable, for 'everything participates in Him' (πάντα αὐτοῦ μετέχει).[62] There does not exist any being that does not

[56] *De beat.*, PG 44:1263d. [57] Ibid.

[58] Ibid. ὁ γὰρ τῇ φύσει ἀόρατος, ὁρατὸς ταῖς ἐνεργείαις γίνεται, ἔν τισι τοῖς περὶ αὐτὸν καθορώμενος.

[59] Ibid. 1265a. [60] Ibid. 1265b, 1269c. [61] *DN* 12.4, PG 3: 972b.

[62] *DN* 5.5, PG 3: 820a.

participate in the Beautiful and the Good.[63] But the Beautiful and the Good are strictly speaking not God in His inmost nature. Participation takes place through what I would call the divine activities. Dionysius has several terms denoting these, such as (*a*) *ἔκφανσις* = manifestation, (*b*) *ἐρωτικὴ κίνησις* = movement of love, (*c*) *πρόοδοι* = processions, and (*d*) *δυνάμεις* = powers.[64]

The term procession or processions seems to be the most common one. Through the processions God is the cause of being, life, wisdom, etc.[65] The processions are powers (*δυνάμεις*), which deify, cause being, life, and wisdom, Dionysius says.[66] But the processions are not different things, many causes or divinities. They are the universal Providence of the one God. They reveal the divine Providence in its more general or more specific aspects.[67] Dionysius wants to emphasize that all processions are ultimately *a unitary manifestation* of the one God. The processions may still be considered in a hierarchic 'movement' downwards, with Goodness on the highest level, followed by Being, Life, and finally Wisdom. But this list is not complete. A lot of processions or 'divine names' may be taken in addition. We have already seen that St Maximus is inspired by some of these ideas, and we will look further at this in the next section.

It is interesting to see that Dionysius tries to make a systematization, a tendency that could on the one hand be seen as lending weight to his doctrine, because somehow his theory becomes thereby easier to understand. On the other hand, the systematization and hierarchization, which generally are a typical feature of Dionysian thought, may give rise to an objection against him. The intelligible cosmos of the Areopagite is usually considered to be strongly coloured by Platonism, and his processions could be viewed as some kind of beings 'between' God and creatures. This is not necessarily wrong in itself if the processions are not hypostasized as a reality of their own. Dionysius is quite explicit on this, when he says that he does not intend to introduce several divinities. The whole

[63] *DN* 4.7, PG 3: 704b: *καὶ οὐκ ἔστι τι τῶν ὄντων, ὃ μὴ μετέχει τοῦ καλοῦ καὶ ἀγαθοῦ.*

[64] All in *DN*, PG 3: (*a*) 4.14, 712c; 2.4, 640d; (*b*) 4.14, 712c; (*c*) 5.1, 816b, 4.14, 712c; 2.4, 640d—used in several places; (*d*) 2.7, 645a.

[65] *DN* 5.2, PG 3: 816c.

[66] *DN* 2.7, PG 3: 645a.

[67] *DN* 5.2, PG 3: 816c–817a.

hierarchy is in the deepest sense just one thing, the manifestation of the infinite and active divine providence. The Dionysian doctrine of processions may easily be confused with Neoplatonic metaphysics, but is in effect something different. The Areopagite has abolished the Neoplatonic hierarchy of beings between the One and the created world. Dionysius, one could suggest, undertakes a Christian revision of Neoplatonic metaphysics when he wants to show that the created world does not participate in God by way of mediating hypostases, but participates directly in God's own processions, i.e. activities.

Here I may be faced with an objection, because the word activity is not documented in the passages I refer to above. To this I will answer that even if the term is not mentioned, it is quite apparent that Dionysian thought is in effect a further development along the same lines as the doctrine of participation in God according to St Gregory of Nyssa. The divine 'names' of Goodness, Being, Life, and Wisdom in Dionysius are divine activities through which God deals with the world.

To sum up. Why do theologians before St Maximus distinguish between God's essence and activities? One can give two answers to this question. First, in the anti-Eunomian polemics of the Cappadocians the distinction was made for epistemological reasons. Second, the distinction is implied by the doctrines of participation in God and the deification of the human being. If the doctrine of deification is not to be taken figuratively or to lead to pantheism, then one has to make a distinction of this kind. But, as I have said above, the distinction is not thematized in the theologians we have investigated.

What then *is* a divine activity considered the second way? This is a difficult question to answer. It seems to be a divine manifestation 'outside of' God's essence. But if that is so, is not the activity to be explained economically and not theologically? What I mean is that it is related to God's creative and redemptive work, and not to the divine being in relation to itself. However, I do not believe that this is all. The concept of activity can be applied to the divine being as well. In that case the original character of the activity must be conceived as something like God's eternal manifestation of Himself through His three hypostases. The terminology of activity of the essence and activity out of the essence may give us a way at least to talk about the difference between the activity as a theological and as an economical matter. As the activity of the essence, it is intransitive and remains with God. As activity out of the essence, it becomes transitive and manifested economically *ad extra*. One thing is at least sure, the activity as God's act is not substantialized or hypostasized as a separate entity. We should look at it as some kind expression or manifestation of the divine being.

III. ESSENCE AND ACTIVITY ACCORDING TO ST MAXIMUS

When we now return to Maximus, I am convinced that the distinction between what I have called God's essence and activities is a part of his philosophy. We find the activities under different names, as we shall see.

In *Cap. gnost.* Maximus distinguishes between two kinds of divine works (*ἔργα*). These are 'the works He began to create' (*τὰ ἔργα ὧν ἤρξατο ποιῆσαι*) and 'the works He did not begin to create' (*τὰ ἔργα ὧν οὐκ ἤρξατο ποιῆσαι*).[68] The term divine works could normally be expected to mean 'creatures'. If we look up the term in Lampe (1989), this expectation is confirmed. God's works are His creatures, but the salvation of man and the works of the Father performed by the Son are also listed as God's works. However, it is quite obvious that it would be wrong to understand divine works as nothing else than creatures in the present context. Further, it generally seems reasonable to think that a 'work' is something *different* from the one who performs the work. There is then a difference between the worker and his work. But we must understand this difference in accordance with the character of the worker and the work in question. A difference could be seen to mean different things depending on the context.

Now, one kind of work is the works which God began to create. They have a temporal beginning (*χρονικῶς ἠργμένα*).[69] They are defined as 'all the participating beings' (*πάντα τὰ ὄντα μετέχοντα*). These beings, which did not always exist, are to be understood as creatures in the proper sense. Such beings are different from God

[68] *Cap. gnost.* 1.48, PG 90: 1100c. [69] Ibid.

because they were brought from non-being to being by Him. They are not divine by nature. Between these beings and their Creator there is, strictly speaking, a basic gap. Their dependence on the divine cause is expressed in the terminology of participation.

The other kind of divine works, however, is something quite different. The works that God did not begin to create could not be understood as creatures. They are defined as 'the participated beings' (τὰ ὄντα μεθεκτά). If these beings are something 'other' than God, they could not be so in the same sense as creatures are. It seems to be a reasonable interpretation to understand the difference between the participated beings and God as a distinction between the divine essence and its activities. These divine works then, are really divine, but they are not God as He is according to His inmost nature. If we read *Cap. gnost.* 1. 47 in connection with 1. 48 it seems a quite reasonable interpretation to hold that the divine works without beginning (1. 48) are collectively identified as the divine activity (ἡ θεία ἐνέργεια, 1. 47).[70]

Being without beginning, these divine works are, of course, not created. In the same way, being without beginning, they are not by nature bound to the institution of creatures. They have their reality independent of God's 'relation' to creatures, even if they play an important part in the institution, preservation and perfection of the created cosmos. The divine works without beginning could be seen as God's eternal manifestation of Himself to Himself *ad intra*. The divine essence itself remains unknowable and the activity of this essence is a mystery. But this divine activity of the essence, as activity out of the essence, according to Maximus, becomes participated (μεθεκτόν) at the appointed 'time', because it is by this power that God gives Himself 'economically' to that which He creates.[71] The divine activity, as God's *works*, become 'participated beings' to creatures when creatures are brought from non-being to being, but also in the preservation of beings and in their fulfilment in a special kind of

[70] Cf. *Cap. gnost.* 1.47–8, PG 90: 1100b–1101a.

[71] Larchet (1998), 56 ff. has made some interesting remarks on the activities in St Maximus' thought. In connection with the *filioque*-problem Larchet argues that the activities are the natural activities of the divine nature common to the three divine persons. He argues that the activities are not only to be understood as a manifestation *ad extra*, but are also an eternal communication *ad intra*.

participation when they reach their final purpose in God. The ontological status of the created world is determined by its participation in the divine activities. In this picture the *logoi*, as we shall see, play an important role.

The terminology of Maximus gives the impression that the activities as beginningless works of God and participated beings are hypostasized as a kind of reality of their own, i.e. as some kind of *beings* established by God. The term 'those that are participated' (τὰ μεθεκτά) was used by Aristotle of the separately existing Platonic Ideas.[72] One can say that the Ideas, as far as they are perfections like goodness, beauty, being, etc., are similar in kind with what Maximus calls divine works. But it is important to remember that for a Christian thinker like Maximus such Ideas could not exist in separation from God.

The beginningless works are not hypostasized as a separate reality. Like Dionysius, Maximus does not want to present a theory of activities as some kind of divine hypostases 'between' God and His creatures. Maximus has a special reason for calling the beginningless works beings (τὰ ὄντα). The reason why he calls them beings is, I believe, something which Dionysius says in *De Divinis nominibus*.[73] Being Itself, according to Dionysius, is more honourable than Life Itself, Wisdom Itself, Likeness to God Itself, etc. *Before* the participating beings can participate in any of the last-mentioned, they have to participate in Being (τὸ εἶναι). 'Before' (πρό) cannot have any connotations to time, but must taken in a logico-ontological sense. The Dionysian idea here is that if a created being has wisdom, it participates in Wisdom, then in Life and finally or primarily in Being. To be wise and to live are certain modes of being. If some thing participates in a certain quality, then it participates in hierarchical order in more and more inclusive qualities. In the end all these qualities are forms of Being, so that the participant always participates in Being as the primary 'activity' participated. To function as principles (ἀρχαί) for created things, they themselves must have being. Maximus thinks in the same way when he talks about participated *beings*. All the other perfections participate, like in Dionysius, in Being which is the primary perfection, in order to function as participable principles.

[72] Cf. *Metaph.* A, 9.990^{b}28–9. [73] *DN* 5.5, PG 3: 820a–b.

Participated beings		Participating beings
ἡ ἀγαθότης	Goodness	good beings
ἡ πᾶσα ζωή	all Life	living beings
ἡ ἀθανασία	Immortality	immortal beings
ἡ ἁπλότης	Simplicity	simple beings
ἡ ἀτρεψία	Immutability	immutable beings
ἡ ἀπειρία	Infinity	infinite beings

Figure 5

The first list of divine works (participated beings) and their participants is given in Figure 5.[74] Maximus does not consider this list as complete. He would like to include as many kinds of being as may be 'essentially contemplated around Him' (περὶ αὐτὸν οὐσιωδῶς θεωρεῖται).[75] Among the participated beings Maximus emphasizes the fundamental character of Goodness, because he speaks about Goodness and all that it implies, such as Life, Immortality, etc. Maximus, so far, seems to be in agreement with Dionysius, who quite explicitly gives primacy to Goodness.[76] According to Dionysius, even if Being plays an important role, it is not the basic activity after all. The Good, on the other hand, tells about all the processions (πρόοδοι), and the divine name 'the Good' reveals the totality of the divine Providence, while the other names (i.e. Being, Life, Wisdom, etc.) specify different aspects of Providence. If the Maximian expression 'all that Goodness implies' indicates that Maximus follows Dionysius here, then he considers the beginningless works listed above as basically united in the divine Goodness. This is in agreement with Maximus' calling these divine works '*the* divine activity' in the singular.[77] All the different elements are brought together in a systematic unity.

There is, however, a difference between Dionysius and Maximus in this regard. Dionysius would not have called Goodness a participable *being*, because Goodness is more fundamental than Being in his system.[78] When Maximus allows this terminology, it surely indicates

[74] *Cap. gnost.* 1.48, PG 90: 1100c–1101a.
[75] We shall return to the meaning of περὶ αὐτόν below.
[76] *DN* 5.1, PG 3: 816b and 5.2, 817a.
[77] *Cap. gnost.* 1.47, PG 90: 1100b–c: τῆς θείας ... ἐνεργείας.
[78] Cf. *DN* 5.2, PG 3: 816c–817a.

Those that are participated		Those that participate	
αὐτὴ ἡ ἀθαναασία	Immortality Itself	τὰ ἀθάνατα πάντα	all immortal things
αὐτὴ ἡ ζωή	Life Itself	τὰ ζῶντα πάντα	all living things
αὐτὴ ἡ ἁγιότης	Holiness Itself	τὰ ἅγια πάντα	all holy things
αὐτὴ ἡ ἀρετή	Virtue Itself	τὰ ἐνάρετα πάντα	all virtuous things
αὐτὴ ἡ ἀγαθότης	Goodness Itself	τὰ ἀγαθὰ πάντα	all good things
αὐτὴ ἡ ὀντότης	Being Itself	τὰ ὄντα πάντα	all beings

Figure 6

that according to him, *Being* is the most fundamental and inclusive. The basic unity of the divine activity is a unity in Being. For Maximus then, Goodness embraces the other activities and is itself embraced by Being. This interpretation is confirmed by the next enumeration of activities in *Cap. gnost.* In this listing (Fig. 6) the activities are not called participable *beings*, but are just seen as 'those that are participated' (τὰ μεθεκτά) in relation to 'those that participate' (τὰ μετέχοντα). We should also note that Simplicity, Immutability and Infinity are not mentioned, while Holiness and Virtue are included.[79] The use of the pronoun itself (αὐτή) to qualify the different activities is interesting. In Plato this is the way he usually denotes the Ideas. Plato speaks about the beautiful itself (αὐτὸ τὸ καλόν), etc.[80] In the *Republic* he contrasts 'the beautiful itself' with all beautiful things.[81] The same way of speaking is found in Plotinus too, and we may see this as a conventional Platonic denomination for the Ideas. In Maximus the 'itself' should be taken in the sense that what we are dealing with here are not instances of immortality, life, holiness, etc., but the perfect manifestations *ad extra* of God that make creatures immortal, living, holy, etc. The activities are not perfect *entities*, but operative divine perfections which are poured forth for beings to be participated in to the degree willed or defined by God.

In connection with these two lists (Figs. 5 and 6) Maximus says 'God transcends infinitely all beings, both participating and participated'.[82] Everything, Maximus continues, which may be categorized

[79] *Cap. gnost.* 1.49–50, PG 90: 1101a–b.

[80] Cf. *Phaedo* 74a, cf. 65d. *Greater Hippias* 286e, *Republic* 493e, *Symposium* 211d, *Phaedrus* 247d.

[81] *Republic* 493e.

[82] *Cap. gnost.* 1.49, PG 90: 1101a.

as *being*, is a work of God.[83] God, as we understand, is beyond all works, transcends all kinds of being. But even if this is so, God is immanent in the world and is participated in by His creatures. The radical apophatisism of Maximus is confirmed in the introduction to the *Mystagogia*, where he emphasizes God's otherness in regard to all things.[84] By Himself God 'never is nor becomes in any way anything that ever is or becomes in any manner'. Consequently, it is more proper to call Him non-being (τὸ μὴ εἶναι) than super-being (τὸ ὑπερεῖναι).[85] To affirm that God is super-being, is a removal (ἀφαίρεσις) of beings, and the other way round, if we affirm beings, we must remove super-being. Now, what does this mean? I think what he says is that if we affirm 'being' of God, even if it be understood in the sense of 'super-being', then 'being' could not qualify creatures at all. If we turn this idea around, to affirm 'being' of creatures implies that 'being' could not have any sense in relation to God. To express it briefly: if God *is*, then beings *are not*, and if beings *are*, then God *is not*. The term 'being' could not be used in common of God and creatures. God, Maximus concludes, is beyond any cataphatic and apophatic discourse. But even if this is so, we may surprisingly enough properly affirm both being and non-being of God. We shall return to the meaning of this dictum below.

This leaves no room for doubt. According to Maximus, we should distinguish between God as He is in Himself or in His own essence, and the activity out of the essence (to use a Plotinian term, but not its conceptual content!) or, to use a synonym, His economical activity. If the activity out of the essence (*ad extra*) is an aspect of the activity of the essence (*ad intra*), which seems to be implied by the term 'works without a beginning' (that transcend the sphere of the economy), then the distinction between the divine essence and its activity is a permanent ontological *proprium* of the divine being. This distinction

[83] Ibid.: *Πᾶν γὰρ εἴ τι τὸν τοῦ εἶναι λόγον ἔχει κατηγορούμενον, ἔργον Θεοῦ τυγχάνει.*

[84] Cf. *Myst.* intr., PG 91: 664a–c.

[85] Perhaps we should translate τὸ ὑπερεῖναι as 'trans-being' rather than 'super-being'. The 'super' may connote a relative, viz. that something is a being to the most eminent degree. The 'trans', on the other hand, connotes the transcendence of something and indicates a basic gulf.

ἀϊδιότης	Eternity
ἀπειρία	Infinity
ἀοριστία	Indeterminateness
ἀγαθότης	Goodness
σοφία	Wisdom
δύναμις	Power[a]

Figure 7

Note: [a] The δύναμις here is the Power to create, preserve, and judge creatures.

between essence and activities may be proved from other texts too, and we shall investigate some of them.

The activities may be contemplated around Him (περὶ αὐτόν),[86] and Maximus distinguishes between what is around (περί) God and what concerns (κατά) Him. In the *De char.* 4.7 Maximus says that God is knowable in ideas or concepts (θεωρήματα) *about* or *around* Him (περὶ αὐτό), but it is not possible to get knowledge in ideas *concerning* Him (κατ' αὐτό).[87] In the *De char.* 1.100 he says that the mind may seek the *logoi* of what is round His essence, but not the *logoi* of what concern His essence.[88] I believe Berthold is correct when he translates *logoi* as 'principles' in this text. I do not think that *logoi* should be taken in the technical sense of *logoi* of beings here.[89] In the *De char.* 1.100 we have another list (Fig. 7) of what I would understand to be basically activities. By these activities God creates, preserves, and judges creatures. The distinction between what is 'about' or 'around' and what 'concerns' God clearly witness to a distinction between the economical activity of God and what concerns the divine being in itself. The 'around' (περί) is a spatial metaphor with the obvious sense of stressing the distinction between God in Himself and His 'surroundings'.

In the *De char.* we find another list of perfections which must be taken to belong to the beginningless divine works or 'those that are participated' (τὰ μεθεκτά). This time (Fig. 8) the perfections are called divine properties (ἰδιώματα).[90]

[86] *Cap. gnost.* 1.48, PG 90: 1100d.

[87] *De char.* 4.7, PG 90: 1049a–b.

[88] *De char.* 1.100, PG 90: 981d–984a.

[89] Cf. Berthold's translation (1985), 48.

[90] *De char.* 3.25, cf. 3.27, PG 90: 1024b–c and 1025a.

τὸ ὄν	Being
τὸ ἀεὶ ὄν	Eternal Being
ἡ ἀγαθότης	Goodness
ἡ σοφία	Wisdom

Figure 8

σοφός	wise
ἀγαθός	good
δυνατός	powerful
φιλάνθρωπος	compassionate
ἐλεήμων	merciful
μακρόθυμος	long-suffering

Figure 9

An additional list of properties (Fig. 9) is implied in another text from *De Char.*[91] Maximus says that when the human mind attains these properties when it is joined to God for long periods through prayer and love. According to Maximus, these properties are first and foremost divine. Therefore it is primarily God Who is wise, good, etc. Maximus substantivates these properties when we talk about God. Men can be wise, but God is not only wise, He is Wisdom Itself, Goodness Itself, etc.[92] But the divine being itself, Maximus adds, is not to be identified with these properties, because God is 'beyond all these'.

The texts I have analysed above witness to a distinction between God's essence and His activity in Maximus. Three questions remain to be discussed. The first one concerns the reasons why Maximus finds the distinction necessary. The second has to do with unity and plurality in relation to the activities. The third one concerns the distinction and relation between the activities and the *logoi*. The third question sets the task for the last section of the present chapter.

Why does Maximus distinguish between the essence and the activities of God? According to Maximus, the Holy Trinity, in Its own essence, is just Itself. As Itself It is not related economically to

[91] *De char.* 2.52, PG 90: 1001b.

[92] Cf. *Amb.* 7, PG 91: 1081d, cf. *De char* 3.27, PG 90: 1025a.

anything at all.[93] In His being God is free and independent. But God also became the Creator and Redeemer of the cosmos. As such He relates to the world as what He creates and redeems. According to Maximus' thought, the unrelated God relates to the world through the expression of His activity *ad extra*. This activity becomes the participated beings of created beings. In the *Mystagogia* Maximus says that both the names 'being' and 'non-being' may be applied to God, because as 'being' God is seen as the cause of beings, and as 'non-being' we deny of Him the being which any creature may have.[94] According to Maximus it is impossible for anything which is 'other' than God to participate in the divine essence itself. Therefore created natures participate in God as activity manifested *ad extra*. In *Cap. gnost.* Maximus says that the participated beings are implanted in creatures by grace, as some kind of innate power 'which clearly proclaims God's presence in all things'.[95] Connected with this ontological dimension of the relation between God and the world, is of course the epistemological dimension. Because God has created the world as something which participates in His activities, man may receive knowledge of God. Works of God which begin their existence in time carry traces of His uncreated activities. A human being, who follows the steps of spiritual development, may come to recognize God through the correct knowledge of creatures.[96]

Are the activities to be understood as a plurality or a unity? On the preceding pages I have often spoken of 'activities' in the plural, even though I have alluded to the fact that these activities in some way are a unity. In *Cap. gnost.* 1.47 Maximus talks about the divine activity which obviously comprises Goodness, Life, Immortality, Simplicity, Immutability, and Infinity.[97] The term 'the divine activity' indicates that what I have called 'activities' really are one single divine activity. In *Ambiguum* 22 Maximus comments on the question of whether all the divine activity (πᾶσα θεὶα ἐνέργεια) signifies that God by Himself is wholly present in an undivided way in each thing.[98] The activity,

[93] Cf. *Cap. gnost.* 1.7, PG 90: 1085b. Cf. the whole sequence in *Cap. gnost.* 1.1–7, PG 90: 1084a ff.

[94] *Myst.* introd., PG 91: 664b.

[95] *Cap. gnost.* 1.49, PG 90: 1101a.

[96] On spiritual development, cf. Thunberg (1995), ch. 6.

[97] Cf. *Cap. gnost.* 1. 47–48, PG 90: 1100b–1101a.

[98] *Amb.* 22, PG 91: 1257b–c.

as that by which God is present, seems in this text to be understood as an undivided whole in itself. It is, however, convenient to talk about 'activities' in the plural whenever we discuss the different ways the activity is manifested in creatures. We should all the time keep in mind the basic unity, which, as we have seen, is *not* to be identified with Goodness, because the texts tell us that all the aspects, Goodness included, are participated beings.[99] Goodness, therefore, is included in the still more basic character of the activity, which is Being. But Goodness is still not just one of the several aspects of the activity, because it comprises a lot of activity-aspects, and is probably intimately connected with Being. Even if the basic character of the activity is Being, Goodness must be a primordial characteristic of Being *qua* divine activity. I find it probable that what Maximus has in mind (at least economically and conceptually) is a hierarchically arranged activity with Being as basic character, but comprising a lot of other aspects in the descending order of a logical sequence. What I have in mind is illustrated in Figure 10. For the details of the list in Figure 10 and the sequence of the aspects of activity, one should look to the next section. Such a scheme is not explicitly developed in any of Maximus' works, even if the above arrangement is suggested by the texts from the *Cap. gnost.* which we have investigated above. I think we are faced with one of the philosophical idea which Maximus has thought through systematically, but which he has not developed at any length in his writings.

The distinction and relation between the activities and the *logoi* is the topic of the next section.

IV. THE ACTIVITIES AND THE *LOGOI*

To my knowledge, only two scholars have seen the point of a distinction between activites and *logoi*. Lossky makes his comment quite generally and not in direct relation to any text from St Maximus. He says that the ideas should be identified with the will or wills (θελήματα) 'which determine the different modes according to which

[99] *Cap. gnost.* 1. 48 and 49–50, PG 90: 1100c–1101b.

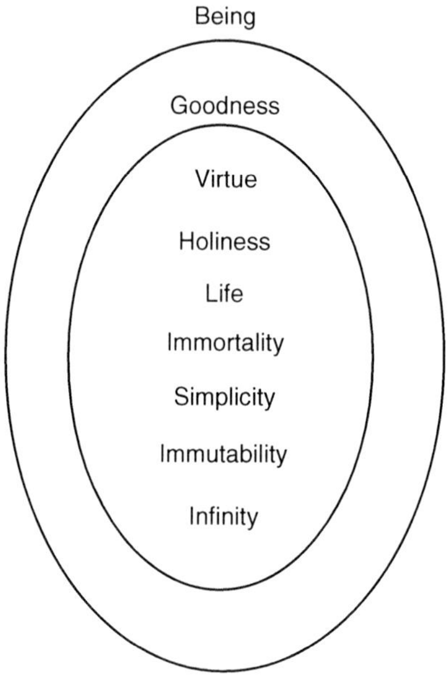

Figure 10 The divine activity

created beings participate in the creative energies'.[100] Lossky does not produce any argument to substantiate this claim, probably because he did not think that it needed any further reasons. Eric Perl follows Lossky when he says the *logoi* are the principles by which creatures participate in the perfections, that is the divine works without beginning.[101]

I agree with both Lossky and Perl, but don't find any clear demonstration in these two authors that the distinction is in fact present in Maximus' philosophy. Does Maximus anywhere explicitly say that activities and *logoi* are different? No, to my knowledge he does not say this in so many words. That the difference is established, however, is quite obvious. If we just take the outcome of the discussion of the *logoi* in Chapter 3 and compare it with what has been brought forward in the present chapter, the picture seems quite clear. The *logoi* are God's intentions through which all creatures receive their generic, specific, and individual essences. The *logoi* are acts of will instituting essence.

[100] Lossky (1973), 95. [101] Perl (1991), 159.

The divine activity, on the other hand, is the manifestation of God's power as Being, Goodness, etc. Even if all the *logoi* are activities of the divine will, all divine activities in the broader sense are not *logoi*.

According to Maximus, divine operation in the broad sense *ad extra* has two aspects. The one aspect is the divine activities in the general sense; the other is specific acts of the divine will (*logoi*). The discussion of the relation between these two aspects brings the whole drama of creation and redemption into view. Maximus says that when God created intelligent beings He communicated (ἐκοινοποίησεν) to them four of the divine properties: Being, Eternal Being, Goodness, and Wisdom.[102] Being and Eternal Being are granted to the essence of the creature, Goodness and Wisdom are granted to its 'gnomic fitness' (ἡ γνωμικὴ ἐπιτηδειότης).[103] Now, the term gnomic fitness poses some difficulties, because the so-called 'gnomic will' (ἡ γνώμη) is usually seen as a characteristic of the sinful condition, and in the present context Maximus seems to speak about man as he was originally created by God. It could be, however, that Maximus uses the term in a rather loose sense to denote the volitive faculty. This should be quite possible, because he remarks in the *Disputation with Pyrrhus* that he has found twenty-eight meanings of the word γνώμη in the Holy Scriptures and in the Fathers.[104]

God gave His four gifts 'so that what He is in His essence the creature may become by participation' (ἵνα ἅπερ ἐστὶν αὐτὸς κατ' οὐσίαν, γίνηται ἡ κτίσις κατὰ μετουσίαν).[105] Even if Maximus speaks about man as originally created, it is obvious that he here distinguishes between the creative and the redemptive orders, i.e. between cosmology and soteriology. The created nature is constituted by participation in the divine Being and Eternal Being, the perfection or salvation of the nature is brought about through participation in Goodness and Wisdom. The nature, as participating, is created in the *image* (κατ' εἰκόνα) of God. The deified nature, as participating, is made in the *likeness* (κατ' ὁμοίωσιν) of God.

God is good and wise by nature (κατὰ φύσιν) and the created intelligent being receives these qualities by grace (κατὰ χάριν). Maximus

[102] *De char.* 3.25, PG 90: 1024b–c.

[103] On 'fitness' (ἐπιτηδειότης), see the treatment at the end of this section.

[104] *Pyrrh.*, PG 91: 312a–c.

[105] *De char.* 3.25, PG 90: 1024b–c.

does not, however, *separate* nature and grace even if he distinguishes between them. The topic of the distinction between nature and grace is rather complex, and it brings us to the heart of the problematic connected with God's transcendence and immanence, and the topic of participation. These are the subjects of the next chapter; however, a few comments must be made now.

Maximus says that participated beings are implanted in creatures by grace (κατὰ χάριν) as a kind of innate power 'which clearly proclaims God's presence in all things'.[106] This is said quite generally and does not seem to be restricted to deification. If creation as well is due to grace, then salvation must consist in a kind of enhancement of the original grace. In the thought of Maximus, every divine act in relation to the cosmos has the mystery of Christ in view. This means that the created status and the redeemed status are not to be separated into two unrelated dimensions. The grace communicated in creation is somehow intensified by the redemptive grace, because redemptive grace brings the participation in God on to higher levels of communication through God's activity. The ontological structure of the created status must be defined carefully with a view to the purpose of all divine activity in relation to the world.

Earlier (Ch. 3 § IV) I focused on the importance of the created will. Even if a creature as a being participates in God, the will plays the decisive role in the creature's status in relation to God. We saw that two modes (τρόποι) are possible, viz. a mode of being that is in accordance with the nature of the creature (κατὰ φύσιν), and a mode of being that is discordant with this nature (παρὰ φύσιν). These two modes could also be designated as 'in accordance with *logos*' and 'discordant with logos' (κατὰ λόγον and παρὰ λόγον). The will decides between these two possibilities. The will therefore, is the 'place' of the tension between different good and evil options.

The drama connected with the terms being, well-being, and eternal being (τὸ εἶναι, τὸ εὖ εἶναι, τὸ ἀεὶ εἶναι) belongs to this picture.[107] The being and the eternal being of the creature are, according to the *De*

[106] *Cap. gnost.* 1.49, PG 90: 1101a.

[107] Cf. *De char.* 3.23–5, PG 90: 1024a–c. The terms τὸ εὖ εἶναι and τὸ ἀεὶ εἶναι are used by Proclus too in his *The Elements of Theology*, cf. prop. 43 (τὸ εὖ εἶναι) and 91 (τὸ ἀεὶ εἶναι). Proclus' use of τὸ ἀεὶ εἶναι is especially interesting as it seems to be connected with ἐπιστροφή, as it is in St Maximus too.

char. 3.25, instituted by the divine gifts of Being and Eternal Being. Well-being, which is equated with participation in Goodness and Wisdom, is given to the will. The first two are beyond the creature's influence, while the well-being of the creature is certainly influenced by its will. One detail of this system, however, is apparently in disagreement with my interpretation in Chapter 3 § IV. Commenting on the triadic structure of the being of man, I connected eternal being with the aspect of deification and the future end of his existence. In such a picture, the eternal being of the creature is something other than a quality of its original essence. The disagreement is, fortunately, only apparent. It has to do with a certain ambiguity in Maximus' use of 'eternal being'.

Eternal being, I just said, is beyond the influence of the creature. But if Maximus is to be consistent, this cannot be quite so. In the *De char.* 3.25 eternal being is connected with the essence. I believe we can understand this to mean that man by nature shall exist forever, but the *quality* of this eternal existence is open to his own influence. His eternal being as a creature has an aspect of potentiality which may be actualized by his own choice. The eternal being of his essence may be actualized either as eternal well-being or as eternal ill-being (ἀεὶ εὖ εἶναι or as ἀεὶ φεῦ εἶναι). The eternal being connected with deification and the final purpose in Chapter 3 §IV above, should be understood as eternal *well*-being.

This interpretation may be compared with what St Maximus says in the *Cap. gnost.* 1.56.[108] The triad (being, well-being, eternal being) is connected with the images of the sixth, seventh, and eighth day of creation. The sixth day betokens the *logos* of the being of beings, because the creation of the cosmos was completed on the 6th day. The seventh day signifies the mode (τρόπος) of well-being, because God rested on the seventh day. The eighth day denotes the mystery of the eternal well-being of beings. I shall not try to interpret the details of this imagery, because it would bring us too far away from our issue. However, the image of the three days is connected with Maximus' threefold scheme of spiritual development. What seems obvious in the text from the *Cap. gnost.* 1.56 is that the eternal well-being of man has to do with the fulfilment of his existence in deification.

[108] *Cap. gnost.* 1.56, PG 90: 1104c.

According to Thunberg, eternal being is usually found to be identical with eternal well-being in Maximus.[109]

How is this participation which constitutes the being, well-being, and eternal (well-) being of the rational creature brought about? Here the terms *logos* of being, *logos* of well-being and *logos* of eternal (well-) being should be taken into consideration.[110] I interpret these *logoi* as *the principles by which* creatures participate in God, i.e. in His activities. If we look to the *De char.* 3.23–5 with this idea in mind, we can say that by the *logos* of being the creature participates God's Being, by the *logos* of well-being the creature participates in Goodness and Wisdom, and by the *logos* of eternal (well-) being the creature participates in God's Eternal Being.

Why does Maximus distinguish between the activities and the *logoi*? At the end of the preceding section I discussed the unity of the divine activity. If the relation of the creatures to God is understood as involving participation, this participation has to be regulated or modified according to certain principles. If there is no such modification, then it is difficult to understand why the creature does not participate in the fullness of the activity at once. It would, consequently, be difficult ontologically to distinguish between the created and the redeemed status. According to Maximus, beings derive from God through creation, and they participate in the divine activity as beings. But, as experience of a sinful and corruptible world shows, they don't possess this activity fully. It seems reasonable therefore, to understand Maximus as teaching that one aspect of the divine operation *ad extra* is the activity, but the relation between the activity and the creatures in the created and in the redeemed status is regulated by another aspect of God's operation, i.e. by the diverse *logoi*, expressed as acts of wills instituting essences. By the *logoi* God diversifies the possible relations that creatures might have to Him, because through these *logoi* He regulates participation according to nature, according to merit and according to deifying grace. This is an important structure in Maximus' ontology of the cosmic drama which takes place 'between' procession and conversion.

The term *logos* of well-being could seem a bit strange. Why not talk about a *logos* of Goodness and a *logos* of Wisdom? With regard to the

109 Thunberg (1995), 371.

110 Cf. *Amb.* 7, PG 91: 1084b–c; *Amb.* 42, 1348d.

lists of activities given in section III this could be taken further: why not a *logos* of Life, of Immortality, of Simplicity, etc.? If we remember Maximus' doctrine of beginningless works of God the answer is obvious. The basic character of the activities is *Being*, so that Goodness is a participated being (ὂν μεθεκτόν). All the activities are participated beings, and the *logoi* therefore, are *logoi* of different kinds of *being* or *logoi* in accordance with which creatures participate in *Being* and all the aspects of the activity which are included in Being. The term *logos* of well-being will cover it all.

We saw above that Maximus makes a distinction between the creative and the redemptive order, i.e. between nature and grace, without separating them.[111] The *logos* of being is related to Being and the *logos* of well-being is related to Goodness and Wisdom. The *logos* of eternal well-being is related to God's Eternal Being, and, as we shall see below in this section, to other aspects of the activity. Here there emerge three possible relations which the creature might have to the divine activity: (i) to Being; (ii) to Goodness and Wisdom; (iii) to Eternal Being which includes other aspects of the activity as well. The first of these relations we could call 'the relation according to nature', while the second and the third are both 'relations according to grace'. The reason for this distinction is that the rational being, as a creature, participates in Being according to its *logos* of being. The participation in other activities depends on the exercise of will. The voluntary movement of the creature brings it into the stages of spiritual development. The participation in the other activities takes place when the creature moves along the threefold path and confirms the unity between its *logoi* of being and well-being, and finally lives according to its *logos* of eternal well-being. In this way God diversifies the possible relations that the creatures might have to Him.

Here we should return to the lists of activities which were given in section III. Could we establish some kind of logical sequence between the different activities? We must remember at the outset, however, that the activities are many; not in themselves, but economically. They are all aspects of the one, divine activity. In addition it must be said that even if some of these aspects might be arranged in a

[111] I use the expression 'nature and grace' for convenience. We should always remember that, according to St Maximus, nature as well is due to grace.

Being → Goodness → Virtue → Holiness → Life

→ Immortality → Simplicity → Immutability → Infinity

Figure 11

logical sequence economically, a total arrangement remains just a hypothesis. I do not claim that Maximus in any text put the activities of God in the sequence established below. I only claim that such a sequence is suggested from intimations in the texts. It is also important to note that we have no reason to believe that the list is exhaustive.[112] It is with these reservations in mind that we may add the lists together.

A comparison of figures 5 and 6 with the relevant texts discloses that the order in which the activities of these lists are arranged seems to be the same.[113] The difference between the two lists, except for omissions and additions, is primarily that they 'move' in opposite directions. The first one (Fig. 5) stretches from Goodness through Life and Immortality (plus Simplicity and Immutability) to Infinity. The second one (Fig. 6) goes from Immortality through Life (plus Holiness and Virtue) to Goodness and Being. This gives the sequence shown in Figure 11. Further it seems reasonable to relate Eternal Being to Being (see Fig. 8). Wisdom is listed next to Goodness in three lists and these two should perhaps be kept together (Figs. 7, 8, and 9). Power could be the divine Power to create, preserve, and judge creatures, and should consequently be arranged with Being and Goodness (Figs. 7 and 9). Compassion, Mercy, and Long-suffering should perhaps be included in Virtue (Fig. 9). Finally it seems reasonable to take Eternity and Indeterminateness with Infinity (Fig. 7). If all this is added together, we get a list in which the sequence is at least probable: Being and Eternal Being, Goodness and Wisdom, Power, Virtue (including Compassion, Mercy and Long-suffering), Holiness, Life, Immortality, Simplicity, Immutability, and Infinity with Eternity and Indeterminateness.

Since the full sequence is hypothetical, I shall primarily dwell on the sequence given in Figure 11. One could hold that the order in which the activities or the aspects of the one activity is listed,

[112] Cf. *Cap. gnost.* 1.48, PG 90: 1100d. [113] Ibid. 1.48 and 1.50.

could be seen as the *logical* sequence in the 'movement' from God as cause to His creatures. God gives Being, Goodness, Virtue, etc. It is rather commonplace in the philosophical and theological traditions to arrange Being and Goodness as the primary ontological structures of the cosmos. God is the origin of the being and the goodness of created things, but what about the sequence of the other activities? St Maximus' idea could be that a creature must *be* if it is to be good, it must be good to be virtuous, it must be virtuous to be holy and it must be holy to have true life. It must have true life if it is to be immortal (in the proper sense of the word), it must be immortal to become simple, it must be simple to become immutable and it must be immutable to have the quality of infinity.

Here a few comments must be made. It seems reasonable to take Life, as it turns up in the list between Holiness and Immortality, not as the condition for biological life, but as the condition of true life. I would think that the condition for biological life is given through participation in Being, because the essence of a living creature is instituted by its *logos* of being which is the divine intention of the *kind* of being that a creature is going to have. True life, on the other hand, is the kind of life received by the creature when it is in active communion with God. One aspect of this true life is that it is immortal life, i.e. that it consists in participation in the gift of Immortality. Even if all intelligent creatures have immortality as a gift endowed to nature, this does not mean that they all will receive the *true* Immortality in eternal communion with God.[114] To be holy then, is to live the true life, and to live the true life is to have true immortality. But how is this kind of life characterized by simplicity? I think this can be under stood in connection with the realization, on the part of the creature, that 'one thing is needful', to quote Christ's words to Martha.[115] To live the true life, conscious of one's true immortality, radically simplifies the rational creature's intention. In this condition the soul is no longer attached to anything worldly.[116] This simplification of the creature's intention could be connected with the so-called 'natural contemplation' which is the second level of Maximus' scheme of spiritual development.[117] In natural contemplation man is no longer

[114] Cf. *De char.* 3.25, PG 90: 1024b–c. *Amb.* 65, PG 91: 1392d. Those who live *παρὰ φύσιν* changes the possible *ἀεὶ εὖ εἶναι* for *τὸ ἀεὶ φεῦ εἶναι*.

[115] Luke 10: 42.

[116] Cf. *De char.* 1.1 ff., PG 90: 961a ff.

[117] Cf. Thunberg (1995), 343 ff.

under the spell of the passions in relation to the outer aspects of sensible things, rather he contemplates the *logoi* of created beings and views all things as united in the Logos. In this condition the gnomic will no longer wavers over different options to satisfy the longing for pleasure, but is firmly established in its course towards God as the cause of all there is. In Maximus' logic this simple and stable intention is further characterized by immutability (*ἀτρεψία*), because man no longer constantly changes his ways (*τρόποι*) to satisfy this or that appetite for pleasure. Being immutable, man receives his share in the divine Infinity. When man has reached thus far, he lives according to the *logos* of eternal well-being and is deified. In the next chapter I shall return to this topic and to the meaning of the participation in divine Infinity. We shall leave the question of the logical force of the sequence in order to collect some loose ends which have been accumulated over the last few pages.

According to Maximus, there is a distinction between the order of creation and the order of redemption. This could be taken as a clear-cut distinction between nature and grace, but here we have to be careful. The creation of nature is itself from God's grace, and to be instituted as a natural essence is the reception of a gracious gift. Despite this, to be, to be well, and to be eternally well are determined by 'the relation according to nature' and 'the relations according to grace'. To be, to be well, and to be eternally well are brought about by the three 'logical' principles, i.e. the *logos* of being, the *logos* of well-being and the *logos* of eternal well-being. The being of the creature ('the relation according to nature') is instituted as a condition in which it participates in Being in accordance with its *logos*. In this condition the essence of the creature is established as an individual, as a member of a species, as a member of a genus and finally as being 'something' (i.e. as just being an essence) in contrast to being nothing.[118] The creation of the rational creature according to its *logos* of being constitutes the creature as participating in divine Being in a limited degree, i.e. as the *kind* of essence it is. God wants this *kind* of essence to exist, and accordingly it receives its share in Being according to its kind. The creature is created with certain natural characteristics, such as

[118] Cf. Ch. 3 § III above on the distribution of essence.

vegetative, animal, and rational powers.[119] The divine image in man, however, is primarily connected with his rational part, i.e. his mind, reason and spirit (νοῦς, λόγος, and πνεῦμα).[120] An important aspect of the rational being of man is that he was created to be self-moving (αὐτοκίνητος).[121] This self-movement could be further qualified as self-determined movement (ἡ αὐτεξούσιος κίνησις).[122] According to Maximus, this rational quality of self-determination is to be identified as the will (ἡ θέλησις).[123] The will, then, is one of the aspects of the image-character of man.

We have now to connect the image-character, which exists according to the *logos* of being ('the relation according to nature'), with the likeness to God, which is actualized according to the *logos* of well-being and primarily according to the *logos* of eternal well-being ('the relations according to grace'). God's Being, Goodness, Wisdom, etc. could not be separated from each other. The three *logoi* of God's will are to be understood as aspects of a unitary divine purpose. But even so, God diversifies the creature's *possibility of participation* by these *logoi*.

Because of its will the rational creature has received the possibility of uniting and separating the three *logoi* in relation to itself, with the consequence that it modifies its relation to God and participates in God's activities in accordance with this unifying and separating activity.

To get a closer view of this drama we may recall the structure of man's being which was analysed in Chapter 3 § IV. By his *logos* of being man is constituted as an essence which joins in the triadic structure of essence–potentiality–activity. The *essence* of man consists in his natural powers, powers of the kind I mentioned above. This essence is the origin of a *potentiality*, which is the capacity of man to effect a change in himself, and to enter into a certain condition. The potentiality is defined as an essential movement, and this is the

119 Cf. the simple enumeration of powers in *De char.* 3.32, PG 90: 1028a; *Pyrrh.*, PG 91: 301a. Cf. *De char.* 3.30–2, PG 90: 1025d–1027a; 4.44, PG 90: 1057b; *LA* 19 ff., PG 90: 925d ff. for a more complex picture. See the detailed analyses in Thunberg (1995), ch. 4, 169 ff.

120 *Amb.* 10, PG 91: 1196a. Cf. Thunberg (1995), 117 ff. for a more comprehensive analysis.

121 *Cap. gnost.* 1.11, PG 90: 1088a.

122 *Pyrrh.*, PG 91: 301b.

123 Ibid., PG 91: 304b–d.

execution of the power of self-movement or self-determinate movement mentioned above. Man is a rational essence with a potentiality to move according to his *will*. Now, this potentiality is the condition of freedom, and Maximus says that each soul deliberately (κατὰ πρόθεσιν) either chooses honour or accepts dishonour by its own works.[124] This means that by his works man may either *unite* the *logos* of being with the *logos* of well-being and the *logos* of eternal well-being, or *separate* the *logos* of being from the other two. These two possible ways of living are the modes (τρόποι) which are in accordance with nature and discordant with nature respectively, or, viewed in relation to the basic principles, in accordance with *logos* and discordant with *logos*.

The human essence exists as such by its participation in the divine Being according to its *logos* of being. The *natural* constitution of man, however, includes more than the rational and volitive powers. According to Maximus, as we have seen, it includes the virtues as well. In the *Disputation with Pyrrhus* this is clearly shown:

PYRRHUS: Virtues, then, are natural?
MAXIMUS: Yes, they are natural.
PYRRHUS: If they are natural, why do they not exist in all men equally, since men have the same nature?
MAXIMUS: But they do exist in all men because of the same nature.[125]

The virtues are not acquired by ascetical struggle, they are not introduced from the outside, but inhere in us from the creation of our nature, Maximus says.[126] How shall we understand this since it seems to mean that man *by nature* participates in God's Goodness and Virtue? Is it not a human task to unite the *logos* of being with the *logos* of well-being, which brings in an additional divine gift? Here we should remember that according to God's scheme, Being and Goodness are not separated, and the three *logoi* are the expression of a unitary divine purpose. Although it is difficult to get a clear understanding of the logic of this complex, the following interpretation could offer a solution: Maximus is convinced that the rational creature has received the virtues *not* as actualized and *not* as a participant in Goodness. The creature has received the virtues according to its

[124] *Cap. gnost.* 1.11, PG 90: 1088a.
[125] *Pyrrh.*, PG 91: 309b.
[126] *Pyrrh.*, PG 91: 309b–c; cf. *Amb.* 10, PG 91: 1109a–b.

logos of being. This means that the virtues are to be understood as a *potentiality* of the rational creature, a potentiality which is connected with the powers of its soul. I think Thunberg is correct when he says that 'the capital virtues of each faculty should be regarded as the proper use of that faculty'.[127] This emphasizes once more that nature and grace cannot be separated. Man is able by the exercise of his will to unite or divide the *logoi* and, consequently, the two basic aspects of the divine activity in relation to himself. The decisive point is the mode (τρόπος) of the actualization of his potentiality. This line of interpretation is illustrated by Maximus' next move in the disputation:

> PYRRHUS: Then why is there such inequality [of virtues] in us?
> MAXIMUS: Because we do not practice equally what is natural. Indeed, if we practiced equally [those virtues] natural to us as we were created to do, then one virtue would be exhibited in us all, just as there is one nature [in us all], and that virtue would not admit of a 'more' or 'less'.[128]

Man was not deified at the outset then, but had to actualize his powers in accordance with his *logos* of being. If he had lived in accordance with his nature he would have confirmed the unity between the triad of *logoi*. This means that he would have actualized himself as participating in the divine Goodness and Wisdom, and all that Goodness implies (i.e. Virtue, Holiness, Life, etc.), and Eternal Being. As it is, however, man never took the decisive step from being in the image to be in the likeness. When he was created he had to make a first move by his rational and volitive faculties according to his nature. Instead of speculating how far man moved according to God's intention before he fell, Maximus just says that man made his wrong choice immediately (ἅμα τῷ γενέσθαι).[129]

The divine image in man is not lost in this fall, because man still exists, if not according to his *logos*, then at least because of his *logos*. His being or nature as instituted by a divine *logos* is not destroyed, neither does it disintegrate. However, the particular nature *qua* manifested in the hypostasis suffers disintegration. This means that it is subject to serious disturbances, and the original well-ordered relationship between its natural powers suffers certain displacements.

127 Thunberg (1995), 293.
128 *Pyrrh.*, PG 91: 309b.
129 *Ad Thal.* 61, CCSG 22: 85.

Man does not hold his being securely, but has delivered himself over to the reign of his passions which split his concern in several possible and conflicting relations to the things around him. The stability of the natural will is shaken by the situation created by the passionate life. In this situation room is made for the ontological modification of the will which operates in the insecurity of fallen existence. Led by his passions man continuously tries to lay the foundation of his being in order to escape the destructive forces of illness, weakness, and death. The passions, however, are not stable at all. They shift from one option to another because they seek what will satisfy human desires. In this situation there occurs a natural bewilderment as to what is good or not good. The mode of will, which operates in this condition of ignorance, doubt, and opposition, is the γνώμη, the 'gnomic will'.[130]

The way back, or the conversion (ἐπιστροφή), is founded by the incarnate Logos Who opens up the possibility of once more joining the *logos* of being with the *logos* of well-being and the *logos* of eternal well-being. Even if man brings himself out of harmony with the basic principle of his being, he does not slip out of God's providence. God circumscribes the cosmos instituted by the *logoi*, so that none of His creatures can risk being separated from God and be dissolved into non-being.[131] According to Maximus, the divine providence is established as the cosmic 'logic' laid down by the *logoi*. In *Ambiguum* 10 he defines providence as the divine care for beings, or as the will of God that gives all things suitable direction.[132] The divine Providence comprises both the universal arrangements of the world and the individuals existing within these arrangements. Maximus exhorts us to believe that what happens happens well, even if the reason is beyond our grasp.[133] Maximus' belief that what happens happens well must be connected with the contemplation of the world from the point of view of its 'logical' or natural arrangement. The free acts of rational creatures, which do not conform to the cosmic logic, are estranged from the *logos* of providence. But this does not mean

130 Cf. *Pyrrh.*, PG 91: 308c ff. Cf. Thunberg's analysis (1995), ch. 4B.

131 Cf. Ch. 3 § I and the discussion of *Myst.* ch. 1, PG 91: 668a–b.

132 *Amb.* 10, PG 91: 1189a–b. For the Maximian idea of providence, cf. the whole sequence in *Amb.* 10, 1188c–1193c.

133 *Amb.* 10, PG 91: 1193b.

that sinful creatures are outside of the divine care. Thunberg sums up Maximus' teaching on God's care for man in the fallen condition thus:[134] 'Man was destined to live eternally, but through his choice of temporal, sensible pleasure he called upon himself—according to God's good council—a pain, which introduced into his life a law of death, which—seen from the aspect of the divine purpose—is there to put an end to his destructive escape from his natural goal.'

I think we here are close to an understanding of the so-called *logoi* of providence and judgement.[135] These are not, I suggest, *logoi* that are different from the triad of *logoi* treated in this section. The *logos* of providence is, as I have just said, the 'logic' of the natural arrangement which God has instituted in the cosmos. But this logic, which provides for the creature in its possibility of attaining its goal in God, becomes on the other hand a *logos* of judgement when violated in the wrong mode of hypostatic existence. When man brings himself into disagreement with the cosmic logic a judgement is inflicted on him. His particular being disintegrates and he earns for himself eternal ill-being. If man changes his ways he is brought back to the divine logic. He is not condemned, but acquitted. This may be a possible way to interpret the *logoi* of providence and judgement in relation to the triad of *logoi*.

The eternal purpose of God is to accomplish the mystery of Christ. Only God incarnate can free human beings from the unhappy dialectic of pleasure and pain. The cosmic role of Christ is described in *Ambiguum* 41. Man was created as a microcosm who should act as mediator in the cosmic divisions between uncreated and created nature, intelligible and sensible nature, heaven and earth, paradise and *oikoumene*, male and female. After having outlined the human task, Maximus continues to talk of the fall, which made the person of the Logos become flesh for the sake of a universal restoration.[136]

It is important to remember that the role of the human nature as microcosm and mediator is not abolished by God's redemptive action. Rather it is by becoming *man* that God accomplishes what man was originally destined to do. Neither is the accomplishment

[134] Cf. Thunberg (1995), 159–60, with references.

[135] Cf. *Cap. gnost.* 2. 16, PG 90: 1132b–c. See Thunberg (1995), 66 ff. for a fuller treatment of the concepts of providence and judgement.

[136] *Amb.* 41, PG 91: 1308c–d.

made by Christ destined to exclude the activity of human *persons*. In Christ the human individual is restored to itself, or, Maximus says, 'to God from Whom I received being and toward Whom I am directed, long desirous of well-being'.[137] The work *Mystagogia* teaches that human beings should be integrated and regenerated in Christ through the Church. In this way one contributes to the completion of Christ, Maximus says,[138] and acts as mediator in Him. The way to reunite the triad of *logoi* is opened up in the 'ecclesial' existence. Man may now move along the path of the three levels of spiritual development and may participate in God's activities in accordance with the *logos* of well-being and the *logos* of eternal well-being.

It seems clear that the sequence of the activities, the triad of *logoi*, the threefold structure of the human being (essence–potentiality–activity) and the three stages of spiritual development are interconnected structures.

When man makes his first move in the natural direction (κατὰ φύσιν) his hypostatic being is set on the way to reintegration. The movement of man according to nature, i.e. according to his *logos* of being, brings him first into the *vita practica* (in Greek: πρακτική, πρᾶξις, πρακτικὴ φιλοσοφία, ἀρετή).[139] We shall return to the parallelism of the first two stages later. In the life of ascetic practice man shall achieve two things, viz. victory in his fight against the passions and the development of virtues.[140] Man thus makes the move which confirms the unity between the *logos* of being and the *logos* of well-being. The hypostatic modification of his natural will as gnomic will is brought into the mode of being in accordance with nature.

Now, in connection with this way of reintegration and salvation we encounter the important question of the relation between the works of man and God's grace. Reading some of Maximus' texts in a superficial manner gives the impression that man has to accomplish a great deal by his own powers to be rewarded by God's gracious gifts. In reality though, the picture is quite different. Human activity is of course important, as it generally is in Greek, Patristic authors. But according to Maximus, every step taken by man in the upward direction is met with a divine step towards man. The question is,

[137] *Myst.* ch. 5, PG 91: 676b. [138] Ibid. [139] Cf. Thunberg (1995), 335.
[140] Cf. *Cap. gnost.* 1.30 and 1.77, PG 90: 1093c and 1112b.

however, who took the initiative? There is no doubt as to the answer, it was God.

In *Ambiguum* 7 Maximus says that creatures participate in God proportionally (ἀναλόγως).[141] Participation takes place according to essential and habitual fitness (ἐπιτηδειότης). I have alluded to the idea of 'fitness' earlier in this section. The term ἐπιτηδειότης is introduced into the scientific vocabulary of Late Antiquity as a development of the Aristotelian doctrine of potentiality and actuality. The background is the realization that potentiality is a *necessary* condition for actuality, but not a *sufficient* one.[142] Sambursky illustrates this point from an example given by Sextus Empiricus.[143] Wood is as such potentially consumable by fire. This is a necessary condition for its being actually consumed. But if the 'suitability', as Sambursky translates the Greek word, is absent, the wood will not burn despite its potentiality. How could the suitability be absent? It could be, for instance, if the wood is soaked. So, potentiality and suitability are the necessary and sufficient conditions for the wood to burn in actuality. According to Gersh, ἐπιτηδειότης is a technical term in Neoplatonism. It denotes the patient's suitability or fitness to receive the influence of the divine cause.[144] Dodds distinguishes three usages of the term.[145] The last one is especially interesting in our context: 'Inherent or induced capacity for the reception of a divine influence.' The essential fitness (ἡ οὐσιώδης ἐπιτηδειότης) in *Ambiguum* 7 could be understood as the fitness of a creature's essence to share the divine activity as Being. Thus it would share the divine activity in a natural, cosmological sense, i.e. according to the *logos* of its being. There seems, however, to be a problem involved in this idea. How could there be such a fitness prior to the actual being of a thing? To admit such a fitness prior to the actual being of the creature would impair Maximus' radical doctrine of creation. The 'existence' of a fitness prior to the reception of being would imply that 'otherness' somehow was given as an eternal condition for the creative act. I think that the problem may be solved by conceiving of the fitness as a quality inherent in the divine design or *logos* for the essence of the

141 *Amb.* 7, PG 91: 1080b. 142 Cf. Sambursky (1962), 106.

143 Ibid. 106–7. 144 Gersh (1978), 37–8.

145 Dodds in Proclus, *The Elements of Theology*, 344 f. Cf. § 1 above on ἐπιστροφή in Proclus.

entity which God wanted to call forth from non-being (cf. Dodd's 'inherent capacity'). Then, when man was created his essence was already designed in the divine *logos* as fit for the reception of the gift of Being, and he was given a share in divine Being according to the *logos* of his being. This gracious gift was the condition that made it possible for man to move naturally. The habitual fitness, or 'the fitness formed by habit' (ἡ ἑκτικὴ ἐπιτηδειότης) could be interpreted as the fitness of the created being to share the divine activity according to the *logos* of well-being, when the creature moves correctly according to the redemptive scheme instituted by the economy of salvation.[146] The habitual fitness exists on the condition of the grace of creation, when man was given the possibility to move correctly. The basic idea seems to be that 'only those in a state of grace may receive grace'.

This requires some further comments. In Chapter 3 § 4 I discussed the triadic structure of man. In that context I noted that deification is somehow a 'natural' actualization or consummation of man, yet this 'natural' consummation is nevertheless not within the power of the created being as such. It seems to me that Maximus distinguishes between the possibilities connected with the natural potential and the fitness to receive the divine activity according to the *logos* of well-being and the *logos* of eternal well-being. The potential for the reception of God's activity is the necessary condition for man to enter a certain actuality. This potential can never be lost. But if man moves in discordance with his nature he will not have the *fitness* to receive this activity. Man did not lose his potential in the fall, but because of man's own activity it was mutilated and rendered impotent as an *actual* capacity to receive. In this way the fitness was lost. In His incarnational dispensation the Logos renewed the capacity which had been mutilated in man's fall. In his ecclesial existence man may achieve a new beginning. Man makes himself fit by habituation when he moves in conformity with the requirements of his being. The movement of man in the *vita practica* is step for step met with a strengthening of the divine presence in man. The virtues are natural, but as 'natural' they are not just 'human'. Their essence is rather the Logos of God.[147] Every man who participates in virtue with a steadfast *habitus*

[146] *Amb.* 7, PG 91: 1080b–c.

[147] *Amb.* 7, PG 91: 1081c ff.

participates in God Who is the essence of the virtues. To develop the virtues one has to cultivate 'the natural seed for the good', which is already present in the human essence. The first move of man in the proper direction is met with an 'incarnation' of God, the Logos, in his virtues. Along this way man comes to participate God as Virtue.[148] We have already seen that well-being consists in the participation in Goodness and all that Goodness implies, including Virtue. In his view of salvation Maximus reckons with both the divine and the human factors. Without the free movement of man nothing will be achieved, but the movement of God towards man is necessary if salvation is to be possible. The process of salvation is a co-operation (συνέργεια) between the two.

The next step on the threefold way is the *vita contemplativa* (in Greek: γνῶσις, φυσικὴ θεωρία, φυσικὴ φιλοσοφία, θεωρία γνωστική).[149] First it is important to note that a condition for arriving in this kind of contemplation is that one has achieved *detachment* (ἀπάθεια) through the practical way.[150] Now, detachment is listed in a logical sequence of virtues in *De char.* 1.2–3.[151] Faith in the Lord leads to fear, fear to self-mastery, self-mastery to endurance and long-suffering, endurance to hope, hope to detachment, and detachment to love. Ἀπάθεια is 'freedom from passion', but there is reason to believe that the term in Maximus does not have only negative connotations. Detachment is the condition of the soul from which love springs.[152] As a condition of contemplation it has two characteristics: (i) if the mind is to be able to contemplate beings according to their *logoi*, then the soul must be free from the tyranny of the passions in relation to the outer aspects of things; (ii) if the mind is to be able to contemplate the *logoi* of beings in their unity in the Logos, then love for God must be awakened in the mind and drive it on to its course.[153] Once again Maximus points out that man's movement towards God cannot be achieved by human efforts alone. This is emphasized in *Cap. gnost.* 1.31. God allows Himself to be touched by the soul, and

[148] On the incarnation of the Logos in the virtues, cf. Thunberg (1995), 323 ff.

[149] Cf. Thunberg (1995), 335.

[150] *Cap. gnost.* 1.32 and 77, PG 90: 1096a and 1112b.

[151] PG 90: 961a–b.

[152] Cf. *De char.* 4.91, PG 90: 1069c–d. Cf. Thunberg (1995), 304 ff. on detachment in St Maximus.

[153] Cf. *De char.* 2.56, 4.45, 1.98, 1.86, PG 90: 1001d, 1057c, 981d, 980c.

He raises it up to Himself. The human mind could not apprehend the divine illumination if God did not draw it up—as far as it is possible for the mind to be drawn.[154]

In *vita contemplativa* man moves further according to his *logos* of well-being, and secures the participation in divine Wisdom. The knowledge of God as the centre of all the *logoi* and as universal cause is only possible if God allows Himself to be known through condescension and by raising the human mind up to Himself. Even if man moves in the correct way, the human mind has not by itself the power to contemplate God. God has to draw the mind up to Himself and illumine it.[155] What else could this mean than that God bestows His gift of Wisdom upon man according to the *logos* of well-being.

The first two stages of spiritual development seem to be intimately connected with each other. They somehow run parallel. That they are considered to be somehow unified is confirmed by the fact that they are both related to the *logos* of well-being and are connected with the human capacity for movement according to will. The parallelism is explicitly brought forward in Chapter 5 of the *Mystagogia*.[156] The human soul, according to Maximus in this text, has two main powers—the vital power (ἡ ζωτικὴ δύναμις) and the intellectual power (ἡ νοερά δύναμις). The intellectual power again has two powers, viz. mind (νοῦς) and reason (λόγος). These two powers are the point of departure of two developments which most certainly are seen to run parallel. The one, which is a development of the capacities of reason, is *practical*. The other, which develops the capacities of the mind, is *contemplative*. Even if the two kinds of life run parallel, St Maximus cannot mean that the parallelism is complete in all details. Individual human beings are probably understood to move differently along the paths and to develop according to individual differences.

The last stage in the threefold spiritual development is the *vita mystica* (in Greek: μυστικὴ θεολογία, θεολογικὴ μυσταγογία, θεολογικὴ φιλοσοφία, θεολογικὴ σοφία).[157] As in the case of the two foregoing stages, I shall not discuss this stage extensively either, but only concentrate on what is relevant to the subject of participation according

[154] PG 90: 1093d–1096a.
[155] *Cap. gnost.* 2.83, PG 90: 1164b.
[156] *Myst.* ch. 5, PG 91: 672d ff.
[157] Cf. Thunberg (1995), 335–6.

to the *logos* of eternal well-being. Now, according to this *logos* man participates in the divine Eternal Being, not only potentially, as he does originally, but actually. In addition, at this stage man participates in Infinity.

Even though man by nature has a potential for receiving Eternal Being, this potential must be understood otherwise than the human potential for a virtuous and contemplative life. The potential for virtue is a positive capacity. Man can develop his fitness (fitness by habituation) for receiving God's grace through the movement in accordance with his *logos*. When it comes to deification, man has a potential for it, and this potential is a necessary condition for being deified. But the fitness is beyond his own capacity. Perhaps we could say that the creature is created with a natural potency which at least does not resist deification, even though the creature does not have a positive capacity for it and even lacks the receptive power. The lack of a receptive power (*δύναμις δεκτική*) is overcome by 'the grace of institution' (*ἡ χάρις τῆς θέσεως*).[158] The receptive power is established by God Himself when the creature receives the grace of institution. The activity of the created triad of essence–potentiality–activity is limited to the actualization of the human being according to the *logos* of well-being. In the consummation of the creature according to God's purpose and the *logos* of *eternal* well-being, God suspends (*καταπαύει*) the activity of the creature by the institution of the grace of deification.[159] This is the paradox that the fulfilment of the creature transcends the creaturely capacity without eliminating the created essence. The suspension of the natural activity of the creature does not mean that this activity is abolished. St Maximus speaks about the creature establishing its natural activity within God Himself.

Here we touch upon a central idea in St Maximus' thought which concerns the logic and ontology of divine immanence and transcendence, and the topic of participation. We shall treat this subject in the next chapter.

[158] *Amb.* 20, PG 91: 1237a–b.

[159] Cf. *Cap. gnost.* 1.47, PG 90: 1100b–c.

5

The Concept of Participation

I. INTRODUCTION

It is a basic idea of Christian thought that God is transcendent. When God and the world are spoken of in relation to each other we must say that each is something totally *other* than the other. God is beyond the world. We may speak of God and the world as distinct 'entities', as *ens increata* and *ens creata*. They do not share anything in common. The one entity, the world, is determined by a set of categories which does not characterize the other entity, God, at all. The ontological inadequacy of the foregoing description is, however, obvious, because according to the prologue of the *Mystagogia* if God is a being (*ens*), then creatures are not beings, and if creatures are beings, then God is not being.[1]

On the other hand, it is a basic Christian idea that God has a kind of immanent presence as well. The Incarnation is a primary instance of God's immanence. To quote St Maximus,[2] 'that which is completely unmoved by nature is moved immovably around that which by nature is moved, and God becomes a human being'. In addition we must remember that, according to Maximus, the Logos of God is 'embodied' or 'incarnated' in nature and Scripture as well. Maximus in several places speaks about God's immanence.

In *Cap. gnost.* 1.49, Maximus says that God transcends all that participate and is participated, but still an implanted power proclaims God to be in all things (οἷά τις δύναμις ἔμφυτος, τὸν ἐν πᾶσιν ὄντα Θεὸν διαπρυσίως κηρύττουσα).[3]

1 Cf. *Myst.* prologue, PG 91: 664a–c.
2 *Amb.* 41, PG 91: 1308d. Louth's translation (1996).
3 *Cap. gnost.* 1.49, PG 90: 1101a.

In *Myst.* ch. 1 it is said that God transcends all, but is still all in all (πάντα γὰρ ἐν πᾶσιν ὤν). He is contemplated in the *logoi*, and He is thus understood as the cause, principle, and end of all creation.[4]

In *Myst.* ch. 7 there is talk of the invisible and unknowable presence of the divine cause in all things. This cause holds all things together and 'renders them unmixed and undivided in themselves and in relation to each other'.[5]

In *Amb.* 10 it is stated that God the Logos, who was incarnated in the last times, is an ineffable, supernatural, and divine fire present, as in the bush, in the being of everything that exists (τοῦ ἀρρήτου καὶ ὑπερφυοῦς ὥσπερ θάμνῳ τῇ οὐσίᾳ τῶν ὄντων ἐνυπάρχοντος θείου πυρὸς).[6]

It should be noted that immanence is connected with the Logos, the *logoi* and the idea of God as cause, that it has to do with the divine activity, and, finally, that it is connected with the topic of participation. The following points sum up Maximus' position:

1. In His own being God transcends every relation and is just Himself.
2. As the cause of creatures God is immanent (*a*) as 'incarnated' through the *logoi* and (*b*) as participated through the activities.
3. As the Saviour of the world God became immanent as incarnated historically.
4. Incarnation (embodiment) is the ontological condition of participation both cosmologically and soteriologically.
5. It is the same ontological logic, which governs the relation between the uncreated and the created being in incarnation as well as in participation. This is the logic of the four Chalcedonian adverbs, cf. § III below.

Maximus' concept of participation has not received much attention as an object of investigation. This is strange, because it is

[4] *Myst.* ch. 1, PG 91: 665b.

[5] *Myst.* ch. 7, PG 91: 685b: καθ' ἥν, ὁ καθ' ὅλου καὶ εἷς τρόπος τῆς ἐν ὅλοις ἀφανοῦς καὶ ἀγνώστου παρουσίας τῆς τῶν ὄντων συνεκτικῆς αἰτίας ποικίλως πᾶσιν ἐνυπάρχων, καὶ καθ' ἑαυτὰ καὶ ἐν ἀλλήλοις τὰ ὅλα συνίστησιν ἄφυρτα καὶ ἀδιαίρετα.

[6] *Amb.* 10, PG 91: 1148d.

commonly held that the idea of participation plays an important role in Greek Christian thought. In the words of Meyendorff:

> In the entire Greek patristic tradition, while any notion of *confusion* of natures is rejected, the idea of *communion* with, or participation in God, is attributed to created nature as its ultimate goal, a goal which is also being anticipated here and now in a variety of ways, even before the eschatological fulfilment.[7]

We should remember that Sherwood wrote in 1964 that a study of participation would serve to clarify what he saw as the acutest problem in Byzantine theology, viz. 'the relation of the finite to the infinite, of the created to the uncreated, not so much in the moment of creation as in the moment of deification'.[8] However, I shall not commit myself to deification only, but will focus on creation as well, because the cosmological and the soteriological dimensions belong together. That participation plays an important role in Greek and Byzantine theologians is commonly accepted among Patristic scholars. It is, however, somewhat frustrating to see that scholars often use the terminology of participation as if a readily accessible and well-defined concept was known to everyone. Almost nobody seems to be aware of any problems connected with the idea. This is seen even in Balás well-known *ΜΕΤΟΥΣΙΑ ΘΕΟΥ*, in which it is hard to find out whether the author actually *defines* the concept or not. I consider this a major weakness with Balás', in other regards, stimulating study.

Given the fact that many scholars acknowledge the important role of participation in the Greek Fathers, it comes as a surprise that Larchet seems to deny that there is any concept of participation in Maximus. According to Larchet, Maximus did not develop a precise doctrine of participation even though he occasionally used the terminology.[9] I shall try to substantiate my disagreement with this point of view and to define Maximus' concept as accurately as possible.

The present topic is both difficult to handle and also obscure. To begin with I will comment on the semantics of some important terms from the vocabulary of participation. This immediately brings one of the problems connected with the idea of participation into view.

7 Meyendorff (1987), 21.
8 Sherwood (1964), 435.
9 Larchet (1996), 600–1.

τὸ μεθεκτόν — the something which is portioned out to be shared by receivers—this implies some sort of division of a participated 'substance'

τὸ μετέχον — the receiver of a portion of a (divided and) shared 'substance', the something which is *having* (as a *habitus*, ἕξις) the portion or its share in what is participated

μετέχω — the act of receiving or *having* (as a *habitus*, ἕξις) a portion of something shared

μέθεξις — the condition of *having* (as a *habitus*, ἕξις) a portion of something shared

μετουσία — the condition of sharing the essence of something, or the condition in which the sharing of the part of the essence shared becomes an essential feature of the recipient

κοινωνία — the communion or fellowship which something has in something, for instance when somebody shares a bottle of wine or bread

παρουσία — the essential presence of something in something

To this some additional remarks must be made. What is meant by *habitus* here will be commented on in section IV. Of the terms μέθεξις and μετουσία, the latter seems to express something stronger than the first. This, however, is not necessarily the case. When μετουσία is used by St Gregory of Nyssa it could not, I believe, mean that a created thing participates in God's essence.[10] The same is the case with Maximus when he uses the term in the *De char.* 3.25.[11] God gives Being and Eternal Being to the human essence, and He gives Goodness and Wisdom to the will 'in order that what He is by essence the creature might become by μετουσία'. I think that the whole logic of Maximus' system precludes this from being taken in the sense that the creature participates in the transcendent divine essence. It is interesting to note that in Proclus the term μετουσία may be used synonymously with the verbal forms μετέχει and μετέχειν.[12] This indicates that μετουσία and μέθεξις are regarded as synonyms by him, and I believe the same is the case in the thought of Gregory and Maximus.

[10] Cf. *De vita Moysis* 2.25. [11] *De char.* 3.25, PG 90: 1024b–c.

[12] Proclus, *The Elements of Theology*, the last lines of prop. 20.

II. THE PROBLEM OF PARTICIPATION

The semantics of the terms listed above have a striking common feature, namely their rather materialistic connotations. As used in philosophical and theological doctrines of principles they should probably be understood as *images* for a certain causal and ontological dependency. Aristotle considered the Platonic μέθεξις as a poetic metaphor.[13] But the materialistic connotations are obviously felt by Plato. In a famous passage in the *Parmenides*, Plato goes through several arguments, which make a problem of the separation and participation of the Ideas in relation to the participants.[14] The material connotations are clearly seen in the first argument. The young Socrates converses with the old Parmenides and says that the whole Idea, which is one, is in each of its participants. Parmenides replies that in that case the one Idea, because it is present as a whole in separate individuals, 'would itself be separate from itself'. Socrates objects that this does not necessarily follow, because the Idea might be likened to the daylight, which is one and the same while present in many places at once, and is not separated from itself. Parmenides, however, says that this would be like a sail spread over many persons, and he poses the question whether the whole sail would be over each person, or a particular part over each. Socrates answers that a particular part would be over each. From this Parmenides draws the conclusion that the Ideas themselves are divisible into parts with the consequences that follow. I am not sure that Socrates should let himself be forced to drop the analogy of the daylight because of Parmenides' argument. Without protest Socrates seems to accept the shift in the argument from the thought that the Idea is something qualitative to conceiving of it in quantitative terms. However, the problem of participation turns out to be the problem of how the many may participate in the one.

In a simple way the *Parmenides* illustrates how problems arise when an immaterial principle is thought to be distributed to sensible things and is participated in by them. Somehow the intelligible principle becomes transformed into a quasi-material 'substance'. It could be argued that the dialogue *Timaeus* offers a solution to the Platonic

[13] *Metaph.* A, 9.991^{a}20 ff. [14] *Parmenides* 130e–135c.

problem of participation when a creative, divine mind is brought into the picture. The demiurgic mind causes things to exist with a certain formal structure, but does not divide and distribute the Idea as if it were a quasi-material 'something'. But in this way the Idea itself is not really participated in.

In Plotinus the Platonic teaching about the making of the cosmos is interpreted in quite a sophisticated way. We may construct the Plotinian model of participation if we bring his doctrine of double activity (ἐνέργεια τῆς οὐσίας, ἐνέργεια ἐκ τῆς οὐσίας) into contact with his thoughts about contemplation and action or production (θεωρία, πρᾶξις, ποίησις).[15] First of all we should note that Plotinus too talks about the participation of the lower in the higher. There are, for instance, some passages in which he speaks about bodies being beautiful by participation (μέθεξις, μετοχή) in a higher Form.[16]

If we combine the principles worked out in *Ennead* 3.8.1–4 with the doctrine of double activity, we get the following picture: as the One remains itself, its 'activity of the essence' is self-contemplation. This contemplation is accompanied by an 'activity out of the essence' as its action or creative activity. This creative activity is established as the next level, the level of the Intellect, because the activity out of the essence, as a *logos*, is turned as self-contemplation towards itself as a derivation of the higher level. Thus, the activity out of the essence of the One is identified as the activity of the essence of the Intellect. This activity of the essence of the Intellect is its self-contemplation and at the same time its self-constitution as an activity deriving from the higher. The same kind of activity is repeated on the level of the Soul and of nature as the lowest level of creativity. I have systematized main points of the scheme in Figure 12. In this system *participation* may be understood as the process of the higher level (i.e. its activity of essence) becoming present at the lower level (as the activity out of the essence). The activity out of the essence of the higher principle, as self-contemplating *logos* on the lower level, is institutive of Form (εἶδος) on the lower level.[17] The Form 'transferred'—to speak figuratively—from the higher to the lower level might be any Form

[15] For contemplation and action, see *Enn.* 3.8.1–4.

[16] Cf. *Enn.* 1.6.1; 5.9.2; 1.6.2.

[17] Cf. *Enn.* 3.8.2.

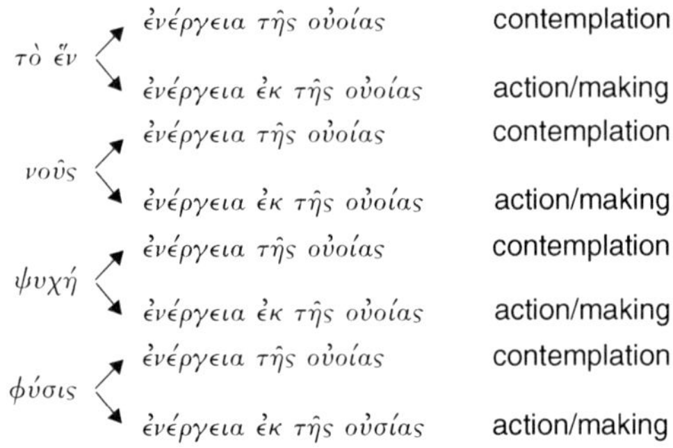

Figure 12.

that is present on the higher level, for instance the Form of beauty in which bodies participate.

This interpretation of participation in Plotinus seems to receive some confirmation from what Dominic O'Meara says in an important article on the topic.[18] He focuses on *Ennead* 6.4–5 in which Plotinus obviously tries to solve the problem of participation. O'Meara says that we should not conceive of the intelligible as 'coming down', being present in and working on the sensible, rather we should envisage the sensible as 'looking up', 'going towards', and being present to the intelligible.[19] However, it seems to me that we should view it both ways, because the activity out of the essence is the condition for the next level to 'look up' through its contemplation of itself as derived from the higher. We should, in other words, think of a twofold presence and this is what O'Meara next admits. The main point of O'Meara's interpretation is to interpret presence on two levels, viz. as 'the total self-presence of the intelligible which is its ontological integrity, and the "presence" of the sensible to the intelligible which is the participation of the sensible in, its causal dependence on, the intelligible'. The sensible 'participates in the same integral totality of Being in which all other sensibles participate: the intelligible is thus "present" as a whole to them all'. O'Meara continues: 'Variation in the presence of the intelligible is explained by Plotinus, not in terms of a fragmentation or parcelling out of the intelligible, but in terms of a

[18] O'Meara (1980). [19] Ibid. 68.

variety in the degree to which sensibles are capable of sharing in the same integral totality.'[20]

The intelligible principle is present as a whole on the lower level through its activity out of the essence, but its *qualified* presence on the lower level is determined by the receptive capacity of the participating entity. The problem now is to account for the gradations of receptive capacity. In Plato's *Timaeus* this is not a major problem, because unordered matter is present as an original principle. In the monistic system of Plotinus, as in Neoplatonic thought in general, there still seems to lurk a dualistic element.

From the interpretation given above and from the citation from O'Meara there emerges an idea of participation as a doctrine of a certain causal relation that is not in any acute way hampered by the materialistic connotations implicit in the vocabulary of participation as such. The reason why I have dwelt so extensively on this problematic in Plotinus is that we find here some clues to the interpretation of Maximus, as we shall see when we come to section IV.

Plotinus uses the terminology of πρόοδος and ἐπιστροφή.[21] The activity out of the essence represents the *procession*, while the self-contemplative activity of the essence, with its self-constitutive effect, represents the *conversion*. The development of this scheme into the elaborate system of triadic and cyclic causation of μονή–πρόοδος–ἐπιστροφή belongs to later Neoplatonism.[22]

What exactly is participated in according to the Plotinian model? It does not seem to me, for instance, that the Form of beauty itself is literally 'transferred' from the higher to the lower level. The 'activity out of the essence' of the Form is an aspect of the activity of its essence, but should also be distinguished as a lower by-product of the primary activity of the Form. The *logos* as self-contemplation institutes the Form at the lower level, but the lower and the higher are not to be identified (cf. the πρόοδος). The Form instituted could be conceived as something 'other' than the Form of beauty itself. Even if the continuity of the system is emphasized, it seems as if the higher Form is received as a kind of replica.

20 Ibid. 70.

21 Cf. *Enn.* 5.5.3; 5.1.7; 5.3.6 (at the end), cf. the opening of 5.3.7.

22 Cf. Gersh (1978), 45 ff.; Dodds' commentary in Proclus, *Elements of Theology*, 220 ff.

Some elements of the Plotinian scheme are further developed by Proclus who uses the triad μονή–πρόοδος–ἐπιστροφή to explain the causal process, which constitutes the lower levels of the cosmic hierarchy. Proclus' concept of participation should be understood in relation to this triad. The effect, Proclus says, remains in its cause (μένει ἐν τῇ αὑτοῦ αἰτίᾳ), proceeds from it (πρόεισιν ἀπ' αὐτῆς), and converts to it (ἐπιστρέφει πρὸς αὐτήν).[23] The *remaining* means that the cause and the effect have the same quality, or better, that the quality, which the effect has, is perfectly present in the cause. The *procession* accounts for the difference between cause and effect, while the *conversion* means that the effect is constituted as an entity with the 'reception' of the quality remaining in the cause.[24] Proclus makes the interesting distinctions between the unparticipated, the participated, and the participant (τὸ ἀμέθεκτον, τὸ μετέχεσθαι δυνάμενον or τὸ μετεχόμενον, τὸ μετέχον).[25] These distinctions are worked into the systems of Dionysius and St Maximus as well.

In Proclus the participated moments of the unparticipated cause are generated by it, and the common quality or the identity between cause and effect on the lower level is secured by the participated moment. In the conversion of the effect to its cause, the effect is constituted in this quality or identity by the presence of the participated moment.[26] If this is to solve the problem of the transcendence and immanence of the higher in relation to the lower in a doctrine of participation, then we must expect Proclus to have a satisfactory explanation of the relation between the participated moments and the higher principle. Unfortunately I lack the knowledge of Proclus' works, which would be necessary to give an accurate philosophical description of this relation. In proposition 23 of *The Elements of Theology*, however, he says that the unparticipated 'will give something of itself' (δώσει τι ἀφ' ἑαυτοῦ), and in proposition 18 he talks about a principle which 'bestows by mere being' (τῷ εἶναι δίδωσι). This could indicate that Proclus seeks a solution to the problem along the Plotinian lines of the activities of and out of the essence. By remaining itself in its own activity the cause gives of itself and its activity becomes transitive. If this is so, the perfection of the

23 *Elements of Theology*, prop. 35.
24 Ibid., props. 30–2.
25 Ibid., props. 23–4.
26 Ibid., props. 23, 24, 140, 182.

manifested reality as participated term and its close relation to its cause, is secured. On the other hand, the participated term is distinguished from the cause, because 'whatever is productive of anything is superior to its product' (prop. 18). The participated term, which next becomes a property of a participant, is secondary to its cause (prop. 23). But if cause and effect are to be distinguished, some element of otherness must be introduced in the effect. This element of otherness must be furnished by the participant, whatever way this is to be understood. Since there are *many* participants, there must be *many* participated moments. I am not sure how Proclus would explain the latter kind of plurality. Could it be that the participants account for the otherness, which make the participated principle into a plurality? According to this view the otherness of the participant becomes the principle of multiplication. If this line of thought is correct, then once again we are confronted with the dualism lurking in the monistic system of Neoplatonism. Perhaps Proclus would try to solve these difficulties by employing the notion of self-constitution.[27] Anyway, creatures, or participants, have an important role to play in the process of creation.[28]

My interpretation of Proclus could be systematized as follows:

The unparticipated cause
(Activity of the essence)

(Activity out of the essence)
The participated moments, which are pluralized
in relation to participants

Participants

The number of participated principles is the same as the number of participants. The participated principle, therefore, plays a role analogous to that of the Aristotelian immanent form.

In its main points the structure of the Proclean concept of participation seems to be the same as in Plotinus. The higher intelligible principle, which is present on the lower level, is, according to Proclus, present as the participated element (τὸ μετέχεσθαι δυνάμενον). This participated element presents the unparticipable element as a whole

[27] Cf. Gersh (1978), 48 ff.

[28] Cf. *Elements of Theology*, prop. 2.

to the participant. On the other hand, the participable element is *the qualified presence* of the unparticipable in accordance with the receptive capacity of the participant.

I mentioned earlier that the Proclean distinctions between unparticipated, participated, and participant are interesting because they recur in the Christian philosophies of Dionysius and Maximus. Dionysius actually calls God 'the unparticipated Cause' (ὁ ἀμέθεκτος αἴτιος) which may be participated in by His πρόοδοι, such as the processions of Goodness, Being, Life, Wisdom, etc.[29] In Dionysius as in Maximus we have here the distinction between God as He is in Himself, free from any relationship to anything else, and God as He manifests Himself in His activities. The ideas that God is present as a whole in His activity and that the reception is qualified according to a delimited receptive capacity are also found in Maximus, as we saw in Chapter 4 § IV.

III. THE LOGIC OF PARTICIPATION

St Maximus saw clearly that according to Christian doctrine the world and God should both exist and be thought of in *togetherness with* and *distinction from* each other, but without confusion and without separation. The doctrine of divine incarnation and of participation required a certain ontological logic. He developed this logic from the acts of the Council of Chalcedon. He earned for himself the epithet 'Confessor' (Ὁμολογητής) for his defence of Chalcedonian orthodoxy against the heresy of Monotheletism. The 'logic' of his system, to which I have alluded several times in the preceding chapters, is conceived from the basic concepts used by the council of Chalcedon to define the relation between the two natures of Christ. In Maximus this logic is not employed as a set of sterile and formalistic thought-forms or rules. It rather becomes a dynamic set of concepts which are used in a creative way, for instance to penetrate into the mystery of God becoming man and man becoming deified. Maximus' 'Chalcedonian' or 'Christological logic' has two important features:

[29] Cf. § III below and *DN* 12.4, PG 3: 972b.

(1) it is a set of concepts showing how to philosophize correctly about the Christian mystery; (2) it is not just pure forms of thought. St Maximus holds that it is a logic of being and denotes structures immanent in the way beings are distinguished and unified in reality. When I say that it has to do with *beings*, I primarily indicate that we are dealing with principles regulating the relation between uncreated and created natures, but, as we will see below, this logic regulates the internal relations between created beings as well.

The *definitio fidei* of Chalcedon criticizes those who have introduced *σύγχυσις* and *κρᾶσις*, and have imagined that the flesh and the divinity of Christ have just a single nature (*μίαν φύσιν*).[30] Further it criticizes the idea that because of the confusion (*τῇ συγχύσει*) between humanity and divinity the divine nature of Christ becomes passible. Against this the council confesses a double consubstantiality of Christ (i.e. with the Father and with us), and pronounces that one and the same Christ is 'acknowledged in two natures which undergo no confusion, no change, no division, no separation; at no point was the difference between the natures taken away through the union, but rather the property of both natures is preserved and comes together into a single person and a single hypostasis'.[31]

The core of this doctrine is the four famous adverbs *ἀσυγχύτως*, *ἀτρέπτως*, *ἀδιαιρέτως*, *ἀχωρίστως* (without confusion, without change, without division, without separation). I shall first make some general comments on these.

(1) *Ἀσυγχύτως*: this term is one of the most important concepts in St Maximus' logic. We shall take a look at its background in the philosophical discussions on different kinds of mixtures. In Stoic literature one distinguished between *παράθεσις*, *μίξις*, *κρᾶσις*, and *σύγχυσις*. These kinds of mixture may be characterized in the following way:

πράθεσις juxtaposition; 'the simple one beside the other in external contact'.[32]

[30] Tanner (1990), 1: 84.40.

[31] Ibid. 1: 86. I have changed Tanner's 'a single subsistent being' to 'a single hypostasis'.

[32] Grillmeier and Hainthaler (1995), 40. Cf. Dörrie (1959), 26–7, on whom Grillmeier builds.

μίξις combination, mingling; 'qualities thought to be corporeal permeate the bodies to be united. Examples of this are fire in iron and light in the air.'[33]

κρᾶσις blending; 'mingling with partial weakening of the original qualities. An example is the pouring together of fluids, which, for instance, produce perfume. A separation is only possible with special means.'[34]

σύγχυσις mixture, confusion; 'the most intense degree of union. From two material elements there results a third with completely new qualities. Examples are medicines... A separation into the component parts is no longer possible.'[35]

In *Opusculum* 18 St Maximus lists twelve types of unions, which are theologically, and Christologically relevant.[36] The heading of the *Opusculum* is *Ὅροι ἑνώσεων*, but the 'definitions' are so short and the examples so scanty that the whole work looks more like a mnemonic list than anything else. Unity according to juxtaposition is exemplified by the togetherness of pieces of wood (boards). The example of blending (*κρᾶσις*) is the blend of liquids, such as wine and water. Mixture (*σύγχυσις*) is exemplified by melting, such as when beeswax and resin are melted together.

St Maximus' definitions of unity according to essence and according to hypostasis are, of course, of major importance. The two definitions of unity according to essence, one at the beginning of the *Opusculum* and one at the end, do not seem to differ from one another. The idea is that when different hypostases (individuals) have the same essence, there is unity according to essence. When entities, which differ by essence, have one hypostasis, then there is unity according to hypostasis. As an example, St Maximus mentions soul and body which are of different essence, but unified in one human hypostasis. One of the important questions, which were discussed among the philosophical schools, was exactly this question of the union between soul and body.

The idea of an unconfused union finds its expression in Christian authors prior to the council of Chalcedon, first and foremost in

[33] Ibid. [34] Ibid. [35] Ibid. [36] *Th. pol.* 18, PG 91: 213a–215a.

chapter 3 of Nemesius of Emesa's *De natura hominis*. Nemesius is important because we know that Maximus studied his work on the nature of man.[37] According to Nemesius, an intelligible substance is capable of uniting with a receptive entity and of remaining unconfused with it while in union.[38] This is the case with the soul in relation to the body: 'Now the soul is united to the body and it is united to it without confusion' (*Καὶ ἥνωται τοίνυν καὶ ἀσυγχύτως ἥνωται τῷ σώματι ἡ ψυχή*).[39] Nemesius thinks that the soul is present in the whole body, but being incorporeal it is not circumscribed to a particular portion of space. It is not contained in the body, as if in a vessel.[40] The presence of the soul to the body is not a presence of location, but the presence of a certain *relation* (ὡς ἐν σχέσει), which could be compared with the presence of God in human beings.[41] According to Nemesius, the relation should be understood as an inclination or a disposition, which the soul has in relation to the body, just as a lover is bound to his beloved. The soul is not locally present, but the relationship, which it has to the body, implies that it works in the body by its activity (ἐνέργεια). The principles Nemesius has advanced to explain the union of soul and body apply, according to himself, even more to the union of the Logos with His manhood.[42] There are, however, problems here because the union between soul and body is described in such a way that one might think that the soul is united to the body only by its powers or activities. This becomes fatal if these forms of thought are transferred to Christology. It should be noted that the fourth anathema from the second council of Constantinople (553) criticizes those who describe the unity between the divine and the human nature of Christ as according to activity and according to relation (σχέσις).[43] But on the view of Nemesius, the union seems to be more intimate than that. He quotes Porphyry who says that an essence may be assumed in the completion of another essence (παραληφθῆναι εἰς συμπλήρωσιν ἑτέρας οὐσίας) in a union which is unconfused. In the background here there seems to be a Plotinian idea: quality, mixed together with matter and quantity, effects the

[37] This is, for instance, demonstrated by Louth (1996), 45, 96, 205, 209, 210, 211.
[38] *De natura hominis*, ch. 3, PG 40: 593b.
[39] Ibid. 596a–b.
[40] Ibid. 597–600.
[41] Ibid. 600a–b.
[42] Ibid. 601a.
[43] Tanner (1990), 1: 114–15.

completion (συμπλήρωσις) of a sensible substance.[44] The resulting union is not just an external one, but is also a completed being. Nor in the case of Christ is the union just external, but between essences of which the one may complete the other and even transform it. The Logos mingles (κιρνᾶται) with human soul and body and remains (μένει) unmixed, unconfused, uncorrupted, and untransformed (ἄμικτος, ἀσύγχυτος, ἀδιάφθορος, ἀμετάβλητος), but at the same time the Logos contributes to the growth and completion of the humanity that was assumed by Him.[45] The nature of this 'mingling' is, unfortunately, not further defined, and the whole picture is not very clear. Nemesius' wavering over different possible ways of explaining the unity of the two natures of Christ suggests that he does not feel quite secure about the correct solution to the Christological problem.[46]

As a key concept of Christology the ἀσύγχυτος ἕνωσις is at the centre of the discussions after 451 (Chalcedon). What emerged as the orthodox way of understanding this kind of union was to take it as a *hypostatic union* (ἕνωσις ὑποστατική), i.e. a *union according to the hypostasis* (καθ' ὑπόστασιν ἕνωσις): the Logos is a divine hypostasis with a divine nature Who assumes a human nature (without a hypostasis of its own) in such a way that the two natures subsist as unified in the one hypostasis of the Logos without any detriment to the natures. An unconfused union *is* a union or a unification of natures or essences, but in such a way that the natures are kept within their definitory marks in the union. The natures are not melted together in such a way that their differences (διαφοραί) disappear and a new entity emerges with a set of properties mixed together from both the original natures. One further point should be noted. The unconfused union, understood as a hypostatic union, has the distinguishing mark that it is not effected by the agency of two natures in themselves. The character of the union is determined by a principle. In Christ this principle is the hypostasis of the Logos (Ch. 3 § IV). In this hypostasis the natures are unified, the hypostasis being the principle of unification.

44 *Enn.* 6.3.15. 45 *De natura hominis*, PG 40: 601b–605a.

46 On problems connected with the Christology of Nemesius, cf. Grillmeier (1975), 389 ff. and (1995), 207 ff.

(2) Ἀτρέπτως: one of the meanings of τρέπω is 'alter' or 'change'. Metaphysically it means that something is altered from one τρόπος to another. *Tropos* here should *not* be understood according to the Maximian *logos–tropos* scheme, but as the natural constitution of something. According to Chalcedonian doctrine, when the natures of Christ are ἄτρεπτοι, this means that even though the natures are hypostatically united, neither His humanity is changed to become divine by nature, nor is His divinity changed to become human. In Maximus' thought, natures are kept inviolate, i.e. they are preserved in their integrity. Christianity is not a mythological doctrine according to which men may be changed into mythological beasts or gods. When fundamental changes occur, as they do for instance in deification, it is always in such a way that the constitutive difference of the original nature is not extinguished, rather it is preserved according to the divine *logos* of the nature.

(3) Ἀδιαιρέτως: that something is ἀδιαίρετος means that it is undivided or even indivisible. In the Chalcedonian definition this could be taken in the sense that the natures of Christ are not divided in the hypostatic union, nor is it possible that the one Christ could be divided. There is an intimate union between the two natures. The idea of undivided or indivisible is well known from philosophical doctrines of Antiquity and Late Antiquity. The classificatory system of the Porphyrian tree is often called a διαίρεσις, a 'division'. An important point with such divisions is that on certain levels natures are understood as ἀδιαίρετος. Individual entities are identical (ταὐτόν, 'the same') in species and therefore they are undivided and indivisible in this regard. Species are identical in genus and are therefore undivided and indivisible in genus. In the background we have, of course, the *logoi* that are principles of the cosmic arrangements, i.e. the lower levels achieve their character as ἀδιαίρετος in relation to a common nature at a higher level of the system. In this we should see an analogy with the paradigmatic instance of the hypostatic union in Christ: two natures are hypostatically united in Christ, and in an analogous way natures are essentially united in common natures at higher levels of the taxomomy. The natures of Christ are not unified, however, as belonging to a higher genus, because there is no common nature between the created and the uncreated essence.

(4) Ἀχωρίστως: ἀχωριστός means 'inseparable'. It could be understood on the basis of the positive χωριστός. In his discussion of the Platonic Ideas, Aristotle denied that they could be χωριστά, i.e. separately existing entities.[47] According to his investigation of being in *Metaphysics* book 7, separateness is a characteristic of οὐσία.[48] This is the only category that can exist separately; all other modes of being are dependent on substance. For Aristotle the concept of separate existence seems to carry the sense of existing *independently* of other things. What the Chalcedonian definition denies is that the two natures of Christ exist separately from one another, i.e. that each of them have independent existence in relation to the other. According to the principles of a natural division (see above) the divine and the human natures differ widely in accordance with the constitutive differences of each nature. Natures that differ widely are normally separate and exist in independence from one another. In Christ, even if His two natures differ, they are undivided hypostatically. Likewise, even if His two natures differ, and therefore should normally have separate and independent existence in relation to one another, they exist as unseparated hypostatically. As a general metaphysical principle, this may be applied in the way that separately and independently existing entities according to the *diastolic* 'movement' are unified in the *systolic* 'movement' of the taxonomic system.

We shall now see how St Maximus applied this logic in his cosmological doctrine. Apart from the unification between the divine and the human nature in the hypostasis of the Logos, this logic regulates the following relationships:

(*i*) *The relations between created essences*: Maximus maintains the important idea of the totality of creation as a *whole*. This whole is understood as οὐσία comprising genera, species, and particulars in a taxonomy characterized as a unity-in-plurality. According to Maximus, each created nature is by a divine decree preserved in its integrity. This preservation does not, however, set aside the specific and generic relationships as real connections between beings. The ontological logic of differentiation and identification is treated above (Ch. 3 § III). Beings are united without melting together in

[47] *Metaph.* M, 9.1085^{b}34 ff.

[48] *Metaph.* Z, 1.1028^{a}31 ff.

confusion. In their unity they are preserved in their own identity without changing their natural (essential) *tropoi* to become some kind of mythological beasts. Further, being naturally divided according to essence, they are not separated in such a way that they lose their natural communion within *οὐσία* as a whole. It does not belong to the nature of creatures to exist in total independence from one another. This interpretation is based on what Maximus says in the *Mystagogia*:[49] 'God realizes this union among the essences of beings (*περὶ τὰς οὐσίας τῶν ὄντων*) without confusing (*ἀσυγχύτως*) them but in lessening and bringing together their distinction, as was shown, in a relationship and union with Himself as cause, principle, and end.' Maximus also says that the causal divine presence (*παρουσίας*) in all things holds them together by His existence in (*ἐνυπάρχων*) them, and He 'renders them unmixed and undivided (*ἄφυρτα καὶ ἀδιαίρετα*) in themselves and in relation to each other'.

(*ii*) *The internal relations between the logoi*: Maximus says in *Ambiguum* 7 that the one Logos is many *logoi* and the many *logoi* are one Logos.[50] Now, in what way are the many *logoi* unified in the Logos? The Areopagite says explicitly that the *logoi* are 'gathered together into one unity within which there is no confusion' (*ἀσύγχυτως ἕνωσιν*).[51] In a related text (see Ch. 3 § 1) Maximus poses the image of the straight lines which proceed from the centre of the circle, and which are seen as 'entirely undivided (*ἀδιαίρετος*) in that position'.[52] The point of this is that the *logoi* are thought to have an undivided existence in God. The question is now whether, on Maximus' view, the *undivided* should be balanced against the *unconfused* in this union. I believe the answer has to be positive. The *logoi* in God are not to be thought of as completely undivided in a confused way. Even though Maximus himself does not explicitly say so, his doctrine should require that the *principles* of the permanent unity-in-plurality of created beings must be preserved in an unconfused union in God. In other words: in His undivided wisdom God knows unconfusedly all that He wills.[53]

49 *Myst.*, PG 91: ch. 1, 668c (Berthold's translation. I have changed Berthold's 'natures of things' to 'essences of beings'. Greek words inserted by me.) and ch. 7, 685b. Cf. *Amb.* 10, PG 91: 1176b–c which witness to the same.

50 *Amb.* 7, PG 91: 1081c.

51 *DN* 5.7, PG 3: 821a–b.

52 *Cap. gnost.* 2.4, PG 90: 1125d–1128a.

53 Cf. *Amb.* 7, PG 91: 1085b.

(*iii*) *The relations between the divine nature and created nature*: I write 'relations' in the plural because the creature is connected with God through the three *logoi* of being, well-being, and eternal well-being. These relations are, in distinction from those commented on above, relations of *participation*. On the level of nature or essence as such uncreated being and created being are neither related nor in communion at all. In Chapter 2 § v I showed that God and creatures cannot exist in metaphysical simultaneity (*ἅμα*) or share a common mode of being. In *Ambiguum* 7 it is said that God transcends *οὐσία* and cannot be defined by any of those categories which delimit created beings.[54] It is obvious that in this connection the communion between God and creatures requires a medium in which to commune, a medium which must be provided by God and be distinguishable from His essence. St Maximus' realistic idea of communion with or participation in God precludes that this medium could be something created. Rather it is several times witnessed in the texts that communion with God is communion with the uncreated itself through the *activity*, i.e. the divine works without beginning or the divine properties.[55] Maximus' concept of participation shall be discussed in the next section, but on the background of what is said above, at least two questions arise: (*a*) When a creature exists through the *logos* of its natural being, what is this *being* by which it exists? What I have in mind is the following: even if the creature exists by participation in God's Being, the question remains whether creaturely being is a created effect of this Being. (*b*) How is the union between the uncreated activity (bestowed through the three *logoi*) and created nature to be understood ontologically?

(*a*) Maximus says that 'being is from Him [i.e. God], but He is not being' (*ἐξ αὐτοῦ γὰρ τὸ εἶναι, ἀλλ' οὐκ αὐτὸ τὸ εἶναι*).[56] He continues: 'Beings have being in a certain way and not simply' (*εἰ δὲ πῶς, ἀλλ' οὐχ ἁπλῶς, ἔχει τὰ ὄντα τὸ εἶναι*), and by having

[54] *Amb.* 7, PG 91: 1081b.

[55] The union and communion between the two natures of Christ is not a nature-union, but a union according to hypostasis. The mutual interpenetration of the activities of the natures into each other deepens this union. In the end, however, the hypostatic union is the mystery *par exellence* because even though a certain ontological structure and logic may be detected and described in this case, an adequate understanding of the union transcends the capacity of any created intellect.

[56] *Amb.* 10, PG 91: 1180d.

being in this way they are limited within place and time conditions.[57] There are several texts in which the idea of a limited share of being as received from God is expressed, for instance: matter cannot possess being from itself (παρ' ἑαυτῆς), being and form (τὸ εἶναι ... καὶ τὸ εἶδος) are given to beings by God.[58]

The question now is what it means to have being ἐξ αὐτοῦ. One interpretation could be that beings have received and hold a *created being* by which they exist. I very much doubt, however, that this is correct. Several things point in another direction. I shall first state my own opinion: the being of beings is *the uncreated divine Being itself*, i.e. the divine activity. If this interpretation is wrong, I wonder why Maximus, immediately after saying that creatures have their being from God, stresses that God is beyond being. To me this means that creatures *are* by sharing the activity, but God in a certain way transcends this activity. Surely, there is no reason to stress that God is beyond creaturely being?

The language of the texts pointed to above have a strong realistic tint. This realism is also found in other texts. In the *De char.* (3.24) it is said that a rational essence (οὐσία) partakes of God (Θεοῦ μετέχει) by its being.[59] This is further explained in 3.25, which states that when creatures were made, God communicated (ἐκοινοποίησεν) to them—among other properties—the divine property (ἰδίωμα) of Being (τὸ ὄν).[60] The verb κοινοποιέω has the ontologically strong sense of making one's own common with another. In connection with God this community requires, as Maximus clearly sees, that God in Himself transcends His Being *qua* activity.[61] Maximus distinguishes between the essence or nature or form or whatness of a creature on the one hand, and its being on the other.[62] The essence is a created essence, brought forth by God, and it receives its presence in the world by sharing as a participant the divine activity as Being.

Would not this give the creature an almost divine character? The answer is no. First we should note that this is not pantheism, because by *nature* the creature does not have a divine character. As a nature it belongs to the created otherness. Further, what is divine, viz. the *Being*

[57] Ibid. [58] Ibid. 1184a. [59] *De char.* 3.24, PG 90: 1024a–b.
[60] Ibid. 3.25, PG 90: 1024b–c. [61] Cf. *De char.* 3.27, PG 90: 1025a.
[62] Cf. *Amb.* 10, PG 91: 1136b and 1181c–1184a.

of this created entity, does not belong to it as something owned by itself. The power to be is not something which belongs to the creature, but is solely in God's hand. Our growth, our health, and our life are not within our power, Maximus says.[63] The Maximian system seems to require that the relation between the uncreated activity and the created nature should be described by the ontological logic discussed in the present section. It would mean that the divine Being *qua* activity is united with the created nature without confusion, without the divinity being changed into something created or the created into divinity, without the elements of the union being divided or separated. Could this be confirmed by textual evidence? It could at least get some indirect confirmation from texts dealing with the idea of participation according to the *logos* of well-being. Let us turn to the second question posed above.

(*b*) What is the result of the union between the divine activity and the created essence? When something is created, its created nature cannot be separated from the being by which it is. They are not two 'things' which are combined at a certain moment. When God wills, by His eternal *logos*, the nature of a created entity is made present in the world. At the outset the creature is not deified, but it is made with a potentiality which includes the possibility of deification if the creature moves according to the divine triad of *logoi*. This movement according to the *logos* of being or according to nature is natural and implies that the initial participation is strengthened. The creature's development according to the second and the third *logoi* brings it further into the divine sphere. The creature is active on the basis of what is given to it at the start, and God gradually institutes the conditions, which makes further movement possible.

When the creature eventually is brought to move in the deified condition it suffers fundamental changes in its character. What, actually, is changed? On this point Maximus is quite explicit: the nature is preserved, its mode of activity is changed. This is primarily seen in Christ, the paradigmatic example. In the *Disputatio cum Pyrrho*. Maximus says: 'For the God of all, having become man, did not alter the *logos* of [human] nature, since if He had, He could not still have been a man without the perfect and immutable *logos* of all [human]

[63] Cf. *Amb*. 10, PG 91:1196c and *Pyrrh*., PG 91: 297a–c.

nature. Rather, He altered the *tropos* [of its existence], that is to say, its conception by means of seed, and the corruption that cometh through birth.'[64] Even though the human nature of Christ is deified, its integrity as human is preserved by its *logos*. In this lies the soteriological guarantee that He is healing and redeeming the total cosmos by being perfectly human. The same idea is expressed in other places of the same work, for instance when Maximus says natural attributes of the two parts of Christ are exchanged in the union, without any change or mixture of the *logoi* of the natures.[65]

An example of this could be Christ's walking on the water. To walk is a human activity, to walk on the unstable element of water shows that a double activity is involved, viz. the human walking and the divine activity which enables Him to actualize a mode of being which transcends what is natural according to the *logos* of nature. The idea is obvious: the divine activity penetrates into the human nature of Christ, but this nature is preserved, secured by its natural *logos* in God. What is changed is the so-called 'mode' of being, i.e. the way in which the human nature exists and executes its natural functions. Before we turn to the idea of mutual interpenetration (περιχώρεσις) we shall try to get a closer look at this *mode*.

When human beings make use of their natural faculties they have a natural activity. They move into an actualized condition in which they act as the kind of creatures they are, that is by the power of being given through their *logos* of being. In the deified condition the creature obtains the divine activity in such a way that its natural actualization is determined with respect to God. This means, I think, that human beings give form to their act by the power of being given by the *logos* of eternal well-being. First, the natural actualization/activity is due to God acting in creatures on the basic level through the *logoi* institutive of the natural condition. Further, the deified actualization/activity is due to God acting in them through a greater measure of the divine activity.[66] What this means is shown, for instance, in *Cap. gnost.* 2.83, where Maximus says 'we have the mind of Christ',

[64] *Pyrrh.*, PG 91: 320c, Farrell's translation. I have changed Farrell's 'principle' to *logos* and his 'mode' to *tropos*.

[65] *Pyrrh.*, PG 91: 296d–297a, cf. 297d–300a and 345d–348a.

[66] Cf. *Cap. gnost.* 1.47, PG 90: 1100b–c.

which means the power of our mind is illuminated and that Christ brings the same activity that He has into it.[67]

This line of thought is developed further in *Cap. gnost.* 2.87–8. When man comes to the perfect rest at the end of the ages, 'he will possess not just a part of the fullness but rather acquire through participation the entire fullness of grace'.[68] Then the creature 'becomes God through participation in divine grace by ceasing from all activities (ἐνεργειῶν) of mind and sense and with them the natural activities (ἐνεργείας) of the body which become Godlike along with it in a participation of deification proper to it. In this state only God shines forth through body and soul when their natural features are transcended in overwhelming glory.'[69] The point is that the natural faculties of the creature do not function only naturally any longer, but the faculties are established in a divine mode of activity because the divine activity becomes the moving force in man. Man walks, but he is able to walk on the unstable element of water. He sees, but strengthened by divine activity he sees beyond what is visible according to the capacity of his created being.[70] He thinks, but his thought transcends discursive reasoning in contemplation of divine realities.

Some interesting consequences of the thoughts developed above might be seen if we dwell for a moment on a rather striking aspect of Maximus' doctrine of deification. In *Cap. gnost.* 2.88 Maximus mentions the 'infinite splendours' (αἱ ἀπεἰραι ἀγλαΐαι) of the 'nourishment' in the deified condition. In deification man participates in Infinity as one of the aspects of the divine activity. Maximus expresses this bold idea in *Ambiguum* 10, where he says the deified person has become without beginning and end (ἄναρχος καὶ ἀτελεύτητος), and that he possesses the divine and eternal life of the indwelling Logos, a life unbounded by death.[71]

That the created being in deification becomes without end is what we should expect, but according to our argument in Chapter 2 § v

[67] Ibid. 2.83, PG 90: 1164a–b. [68] Ibid. 2. 87, PG 90: 1065b–c.

[69] Ibid. 2. 88, PG 90: 1065d–1068b, trans. Berthold (1985), Greek terms inserted by me.

[70] Cf. the Transfiguration of Christ, for instance in Matt. 17: the apostles saw His invisible glory with their bodily eyes.

[71] *Amb.* 10, PG 91: 1144c.

only God is ἄναρχος. However, in receiving its share in the divine according to the *logos* of eternal well-being man comes to participate in the unoriginate divine activity and quite logically he receives the characteristics of this activity. But even if this is 'logical' it seems to be a major paradox, because how could it be possible for a being which surely has a beginning of its existence to become without beginning? Is not this a contradiction? A creature, which has an origin, becomes without origin and remains from then on both originate and unoriginate. However, to defend the logic of the idea one could assert that the creature is not originate and unoriginate in the same respect: according to essence and hypostasis it is originate, according to mode of being it is unoriginate. Even if this kind of existence in its depths must remain a mystery, it at least does not violate the Aristotelian principle of contradiction. This much is at least clear, Maximus' bold idea emphasizes the 'overwhelming glory' of the divine purpose for created beings, and it emphasizes too that all this talk about deification is not figuratively meant. Man becomes interpenetrated by God, and he becomes God even though by himself he is a weak creature. Maximus boldly states that man is made God, *except for identity of essence* (χωρὶς τῆς κατ᾽οὐσίαν ταὐτότητος).[72]

His character as God is neither by his own nature, because as a creature he has his beginning from non-being, nor by participating in God's very nature, but by grace and participation in the divine activity:

> In Christ who is God and the Logos of the Father there dwells in bodily form the complete fullness of deity by essence (ὅλον κατ᾽ οὐσίαν οἰκεῖ τὸ πλήρωμα τῆς θεότητος σωματικῶς); in us the fullness of deity dwells by grace (ἐν ἡμῖν δὲ κατὰ χάριν οἰκεῖ τὸ πλήρωμα τῆς θεότητος) whenever we have formed in ourselves every virtue and wisdom, lacking in no way which is possible to man in the faithful reproduction of the archetype.[73]

St Gregory Palamas repeats the Maximian idea and even strengthens it when he says that those who attain deification 'become thereby uncreated, unoriginate, and indescribable (ἀκτίστους, ἀνάρχους καὶ ἀπεριγράπτους), although in their own nature, they

[72] *Amb.* 41, PG 91: 1308b. [73] *Cap. gnost.* 2.21, PG 90: 1133d.

derive from nothingness'.[74] None of this means that God and man are ontologically identified with each other, but rather that they communicate across the basic gulf between created and uncreated nature. When God became man He remained God, and when man becomes God he remains by nature man. The union between God and man, between man and God is a kind of unconfused union which we have met above as an important motif in Maximus.

There is one last point, belonging to the subject of participation, on which I will comment briefly in this section, viz. the idea of *περιχώρησις*.[75] The *περιχώρησις* doctrine is a beautiful piece of theological reflection. Maximus holds that not only is man deified by the penetration of the fullness of the divine activities into his natural functions; the incarnated God Himself is humanized by the penetration of the activity of the human nature into the divine nature. This doctrine is, of course, of soteriological importance, because it contributes to the idea of the unity of the God-man and therefore sheds light on the redemptive sufferings of the incarnated Logos. On this background it is also a psychologically important idea in that the loving God is understood to receive into His own nature something of what we are, and therefore freely identifies Himself with us.[76]

IV. INCARNATION AND PARTICIPATION ACCORDING TO ST MAXIMUS

The divine causality in relation to the world is seen by St Maximus as a kind of immanent activity. This immanent activity has two aspects, both of which are connected with the incarnational model. The one Logos is the centre of the many *logoi*, and by these *logoi* He brings about what is accomplished in the two aspects of the activity. The one

[74] *The Triads* 3.1.31, trans. by Gendle (1983), 86, Greek inserted by me. Greek text in Grégoire Palamas ed. Meyendorff (1959), 617.7. On this subject, cf. Christou (1982).

[75] For a broader treatment of *περιχώρησις* see Thunberg (1995), 21 ff.; cf. the index (Périchorèse) in Larchet 1996.

[76] Cf. *Amb.* 5, CCSG 48.33.272–34.296 (PG 91: 1057d–1060b); *Pyrrh.*, PG 91: 296d–297a, 345d–348a.

aspect has to do with the *logoi* as defining the essence of creatures; the other one is that the divinely defined creatures are given their share in being in accordance with their kind. According to the first aspect the Logos acts as the formal or exemplary cause of creatures, and according to the second aspect He acts as the efficient cause because He brings what is essentially defined from non-being to being.

The second aspect of the divine activity is also connected with the incarnation motif, but is not to be understood as an incarnation or embodiment in the proper sense. In the second thesis proposed in section I above I pointed to God's *immanence* as incarnated and participated. In the fourth thesis I asserted that *incarnation* is the condition of *participation*. Existing as Himself and acting immanently through the *logoi* as formal cause, the Logos is *the unparticipated cause*. As representatives of the one Logos to the many creatures the *logoi* too are unparticipated. On the other hand, the incarnational activity of the unparticipated cause(s) has the important effect that creatures become participants of God. Acting as efficient cause the Logos with His *logoi* gives being to creatures. When giving, He limits the share in being which this and that kind of essence is to have according to the divine scheme of things. The second aspect of the divine activity is then, the *participation* in the divine activity actualized by the incarnation or embodiment of the Logos-*logoi*. The activity or all the aspects of the activity should be identified as *the participated moments*.

Now, here we have Maximus' version of the Proclean distinctions between the unparticipated, the participated, and the participants. Maximus' version differs from Proclus' doctrine. According to Proclus the unparticipated cause produces the participated moments as participated hypostases (μετεχόμεναι ὑποστάσεις).[77] Interpreted strictly according to the Proclean model, the Logos should be viewed as the cause and the *logoi* as the participated moments. However, the *logoi* are, as we have seen, not hypostases, but the intentions of the Logos or His acts of will which design creatures. The one Logos could not be multiplied into many *logoi* all of which should be participated entities existing 'between' God and His creatures. This idea would be in conflict with the doctrine of the Logos as *one* of the persons of the

[77] *Elements of Theology* prop. 23.

Holy Trinity. The one Logos is not a myriad of hypostases, but He comprises a myriad of acts of will. All these acts of will are a unity in the Logos of God, even though He by them is present in each and every creature as the innermost foundation of their essence. Maximus interprets the nature of things according to the principles of the Incarnation of the Logos and the mystery of Christ. In His historical Incarnation as Jesus Christ, the Logos becomes immanent, but He does not become participated by His human nature, nor does He, as God, participate in His own humanity. On the other hand, the Incarnation makes participation possible. Therefore, the human nature of Christ is deified *by participation* in the divine activity. Maximus sees this glorification and deification as the divine purpose for the whole created world, and he is convinced that the principle, which makes this possible, is the same on the cosmic as on the historical scene. The historical Incarnation is the condition of the universal deification, but preparatory to this is the embodiment of the Logos by the *logoi* which brings creatures to the threshold of the triadic divine logic of being, well-being and eternal well-being. By the *logoi* of these kinds of being creatures may participate in God in proportion to their development in the direction towards God as the final cause of all there is.

This interpretation is confirmed by textual evidence. We shall take a section from *Ambiguum* 7 (1080a–1084d) as starting point. Here Maximus connects several of his basic ideas, such as the doctrines of creation and redemption, and the doctrine of the three *logoi* of being, well-being, and eternal well-being. He comments on the expression by St Gregory the Theologian of men being 'portions of God' and talks about participation according to its cosmological and soteriological dimensions. First we should note that the Logos of God is transcendent, and is not participated in by any being in any manner by anything at all (*οὐδὲ ὑπό τινος οὐδαμῶς καθ' ὁτιοῦν μέτεχεται*).[78] But does St Maximus anywhere, in the present or in other texts, say that the *logoi* too are unparticipated? In *Ambiguum* 7 he does not say so, but in *Ambiguum* 10 I think he does.

The relevant section (PG 91: 1172a–d) is a 'Contemplation of the future world, and of the gulf, of Lazarus, and the bosom of Abraham'.

[78] *Amb.* 7, PG 91: 1081b/c. Cf. *Qu. Dub.* 173, CCSG 10, 120.

It contains information that is important for the relation between incarnation and participation. St Maximus says that the Logos provides all to all who are worthy proportionate with the quality and quantity of each one's virtue. This, I suppose, is a reference to the 'fitness by habituation' (ἡ ἑκτικὴ ἐπιτηδειότης) commented on above (Ch. 4 § IV). The Logos 'divides Himself indivisibly' and 'is not shared out to those who participate in any way whatever' (ἀμερῶς ἑαυτὸν ἐπιμερίζοντα καὶ τοῖς μετέχουσιν οὐδ' ὁπωσοῦν συνδιατεμνόμενον).[79] We should ask what it means that the Logos divides Himself indivisibly. Maximus further says that the Logos is 'paradoxically' present to each of the participants according to worth. I think this is the same idea as the one we find in *Ambiguum* 7 when in continuance of the assertion, mentioned above, that the Logos is not participated in at all, Maximus says that 'the one Logos is many *logoi* and the many *logoi* are one Logos' (πολλοὶ λόγοι ὁ εἷς λόγος ἐστὶ, καὶ εἷς οἱ πολλοί). The one Logos divides Himself, neither by becoming actually divided, nor in the way a Proclean monad divides itself—i.e. as giving rise to several participated entities—but by directing His *logoi* as His acts of will towards the creation of a plurality of essences. In this way He divides Himself in His creative activity in relation to many things and remains Himself, as the personal subject of this creative will, an undivided unity. This is the 'paradoxical' presence. That the Logos presents Himself according to the *worth* of the participants, means that He presents Himself in accordance with the triad of *logoi* related to the threefold scheme of spiritual development. It is important to note here that the Logos Who divides Himself indivisibly is *not* shared out to those who participate. The Logos, even through His *logoi*, is not shared out because He is the unparticipated cause Who presents Himself to each being as their transcendent cause becoming immanent by His creative acts of will.

We return to the text of *Ambiguum* 7. The Logos with His *logoi* is the unparticipated source, but even so, every creature participates in (μετέχει) God proportionally (ἀναλόγως).[80] Participation takes place on different levels of reality, by mind, reason, sensation, vital movement, or generally by essential and habitual fitness, St Maximus says. The essential fitness (ἡ οὐσιώδης ἐπιτηδειότης) could be taken as

[79] *Amb*. 10, PG 91: 1172b–c. [80] *Amb*. 7, PG 91: 1080b.

the fitness of a creature's essence to share the divine activity as Being (Ch. 4 § IV), i.e. to share in the divine activity in the natural, cosmological sense according to the *logos* of being. The habitual fitness (ἡ ἑκτική ἐπιτηδειότης), on the other hand, is the fitness of the created essence to share in the divine activity by the *logoi* of well-being and eternal well-being when the creature moves correctly according to the redemptive scheme of the economy of salvation.[81] The distinction and relation between incarnation and participation in the process of regeneration and deification is also seen in Maximus' interesting idea of God and man as paradigms of one another. To the degree that 'God is humanized to man through love for mankind, so much is man able to be deified to God through love'.[82] Maximus further talks about the power which deifies man to God through love for God, and which humanizes God to man by love for man, so that the other way around God becomes man by man's deification and man becomes God by God's humanization.[83] This parallelism cannot be on the level of nature and mean that man becomes God by nature to the degree that God became man by nature. If that was the case, we should have to admit mythological transformations or, at least, that to the degree the Logos (being a divine person with a divine nature) took upon Himself human nature, a human person (with a human nature) would receive the divine *nature*. This, however, is precluded by Maximus. Another possibility would be that while the Logos took upon Him human nature, a human person would receive the divine *activity*. But in that case the parallelism would not be perfect, and Maximus writes as if it should be understood as perfect. As far as I can see the most plausible solution is the following one: to the degree that the Logos is humanized by receiving the natural activities of the human nature, the human person is deified by the gracious participation in the natural activity of the divine nature. If this is correct, then we may see a distinction between incarnation and participation, the first being the condition for the second.

There are two further texts from the *Ambigua*, which confirm this relation between incarnation and participation; these are *Ambiguum* 10.1144c–1148a and *Ambiguum* 71.1412b–1416a. In the first text we

[81] Ibid. 1080b–c. [82] *Amb.* 10, PG 91: 1113b.
[83] *Amb.* 7, PG 91: 1084c.

find that a rational being becomes without beginning and end when it is deified by possessing the divine and eternal *life* of the indwelling Logos (τὴν θείαν τοῦ ἐνοικήσαντος λόγου καὶ ἀΐδιον [ζωήν]).[84] The second text emphasizes that the Incarnation is necessary if the divine *Goodness* is not to remain unmixed (ἄμικτος) and the divine love for men unshared (ἀκοινώνητον).

On this background, what exactly is Maximus' concept of participation? Does he have a well-defined concept? I believe he has. One could object that there is no text in which he actually says that 'participation means so and so'. I admit this, but even though a philosopher does not explicitly define a particular *term* it does not follow that he does not have a *concept*. On certain conditions it is still possible to detect concepts, even well-defined ones, in key passages of their writings. In the actual case I am convinced that this is possible with Maximus. Many texts contain the vocabulary of participation. Some of these might offer a clue for solving the problem, but I think our search for a solution should not necessarily be limited to passages in which certain *words* are used. Of course, all of this depends on the hermeneutical principle. If we have a rather vague idea of participation, we may naturally be led to texts in which certain words are used. What we get out of such passages could, however, be rather disappointing. If, on the other hand, we have worked with the topic from a historical perspective and have gone through the problems connected with the classical texts, we should be better prepared for our task. It could further happen that we have found—in authors who are more or less in the same philosophical tradition as our author—the contours of a concept which avoids some of the traditional problems. In that case we are made intellectually sensitive to detect formulations in our author, which would make it possible to frame a more certain hypothesis about what exactly our philosopher means whenever he uses the vocabulary of participation.

O'Meara's interpretation of participation in Plotinus suggests a positive angle from which Maximus could be interpreted because there is something recognizable in Maximus' texts. Once more the question of his sources turns up, but it seems to me almost impossible to answer this question concretely. It is obvious that Maximus must

[84] *Amb.* 10, PG 91: 1144c.

have had access to some Neoplatonic writings, but exactly which writings I cannot say. He studied Dionysius the Areopagite, however, and the contours of a concept of participation may be found in this author.[85] In the case of Maximus, O'Meara's definition of the Plotinian concept could be developed a bit further. I believe the following points serve to characterize the Maximian concept of participation:

1. God the Logos with all His *logoi* is the transcendent cause of all creatures. He is not participated in by anything.
2. There is a divine activity which is manifested for the creation and redemption of the world. This activity is the object of participation.
3. Participation should not be thought of as a 'coming down' of the activity to be divided and distributed among the participants.
4. The divine activity presents God as a simple undivided whole to each participant.
5. Creatures are designed by God in the *logoi* in such a way that their nature has a limited degree of receptive power (ἐπιτηδειότης).
6. When God so wills, creatures emerge into the presence of being by the actual reception of the divine activity to the degree delimited by the *logoi*.
7. Creatures receive the power of the undivided activity as Being, Goodness, Wisdom, etc., and thereby become existent, good, wise, etc., according to their receptive potential.
8. The created essence does not exist by a created being (*esse*), but by the reception of God's activity (ἐνέργεια) as Being. Likewise, the created being is not deified by the reception of a created perfection, but by the reception of God's activity as Eternal Being. However, according to St Maximus, the uncreated and the created are kept within their proper spheres according to the ontological 'logic' explained above.

If we take the vocabulary of participation literally, we should expect the intelligible principle to be divided and shared out as some kind of

[85] Esp. *DN* ch. 4.

quasi-material substance. The problem about this is already felt, as we have seen, in the *Parmenides* of Plato. There is no passage in Maximus, known to me, in which he expresses such a naïve view. If we set aside the materialistic connotations of the vocabulary it could be taken as a conventional tool for expressing a subtle metaphysical doctrine. The main point in Maximus' concept is the idea that lower reality could neither exist nor be glorified if the cause did not make itself present as that by which this lower reality has *being* in a certain mode (τρόπος). The other side of this is that the creature emerges from not-being-present to be present and further on to be present in more advanced ways. The modes of these kinds of presence are qualified by the *logoi* which define the creatures as essentially determined hypostases (or individual beings) suitable to exist in just these modes. The most important text that I have found which gives the contours of such a concept of participation is from *Ambiguum* 22.[86]

Maximus is commenting on a difficult passage from *The second theological oration* of St Gregory the Theologian:[87] 'Now the subject of God is more hard to come at, in proportion as it is more perfect than any other, and is open to objections, and the solutions of them are more laborious.' In his comments Maximus immediately turns to one important issue, viz. the problem of how God, through His *logoi* and His activities, is related to the plurality of the created world. I don't think that the text from the Theologian is just grasped by Maximus as an opportunity to develop his own ideas. There are certain cosmological topics mentioned in the oration which are related to the Maximian metaphysics.[88]

Creatures are many, Maximus says, and as many they differ according to differences introduced by God through the *logoi*.[89] Now, if creatures differ, one must understand that the *logoi* by which the creaturely essences differ, differ as well. From this we know there are many different *logoi*. Then he introduces the problem: it belongs to the nature of the senses to grasp the sensible as many and differing sensations. Likewise, the mind (νοῦς), when it grasps the *logoi* which

[86] PG 91: 1256d–1257c.

[87] PG 36: 53b. Translation from NPNF 7: 296.

[88] e.g. in 48a St Gregory speaks of a *logos* being implanted in the world, in accordance with which the world is moved and administered.

[89] The first part of the argument is contained in PG 91: 1256d–1257a.

are in beings, makes them many and infinite. The problem is that the mind (or the intellect) lacks the power to gain a scientific understanding of reality, because it does not grasp exactly how God is related to the manifold of the world. The thinker receives manifold impressions through his senses, he grasps a manifold of essences by contemplating beings, and he reasons to the existence of a manifold of principles (*logoi*). But how should he conceive this manifold in relation to the simple being of God? God, Maximus says, is nothing of what is, He is all of it and He transcends it all (*ὁ μηδὲν ὢν τῶν ὄντων ἀληθῶς καὶ πάντα κυρίως ὢν καὶ ὑπὲρ πάντα Θεός*).

How could God be both different from and identical with beings? The solution is to distinguish, as I have done in Chapter 4, between the divine essence and its activity. God is not different and identical in relation to the world in the same respect. He differs radically according to essence and is identical according to activity. However, when we speak of 'identity' (cf. above 'He is all of it'), we are not entitled to think that the divine activity is literally *the same* as the world of created essences. The point could rather be that the activity is *that by which beings are*, and without this activity of Being, nothing at all can be.

In the next step of the argument the problem of the simplicity of God and the manifold of the world is exposed in greater detail.[90] The totality of the divine activity (*πᾶσα θεία . . . ἐνέργεια*) by itself signifies God's being present as a whole in an undivided manner in each thing (*τὸν Θεὸν ἀμερῶς ὅλον δι' ἑαυτῆς ἐν ἑκάστῳ*). What the interpreter would like to know is how God, by the activity, is present as a whole in all creatures in a common way (*κοινῶς*) and in each creature in a special way (*ἰδιαζόντως*), since He is both without division and without being imparted (*ἀμερῶς τε καὶ ἀμερίστως*). He suffers neither distribution nor contraction, but still He is all in all (*ἀλλὰ πάντα ἐν πᾶσίν ἐστιν*), Maximus says.

These are the main lines of Maximus' thought as it is developed in the text. What conclusions can be drawn from this? First, we should note that God's presence is a presence by way of the activity. Second, Maximus makes three points about this presence: (1) God is present as a whole in all creatures in a common way. (2) He is present as a

[90] From 1257a–c.

whole in an undivided way in each thing. (3) He is present as a whole in each creature in a special way, i.e. in a way proper to each entity.

The first point could be explained if we remember the basic ontological structure of *οὐσία*, which includes all genera, species, and particulars. *Οὐσία* includes the totality of creation in the way that a whole contains its parts. It is the highest universal, not in the abstract sense, but in the concrete sense of a unifying reality immanent in the order of created beings. God can be present as a whole in all creatures, i.e. in a way common to all, if He is present in this basic structure of *οὐσία*. We saw earlier that the activities of God are really one, simple divine activity (Ch. 4 § III). It is as received by creatures that the activity is diversified, i.e. in its effects. By actualizing this and that in the receivers or the participants we come to distinguish hierarchically between aspects of the activity. On this background we could say that point (1) concerns the reception of the activity as Being. By the presence of the activity as Being God is present as a whole in a common way in the all-inclusive *οὐσία* as the basic fact of created entities.

Further, the activity is not only present to the whole of creatures, but also to each individual being contained in the whole. This presence is twofold. Point (2) concerns God's presence as a whole in an undivided way to each creature. This, I believe, is just an aspect of the first-mentioned kind of presence. The total presence in a common way to *all* creatures includes that God is present to *each* in the same way. The reason is that each created essence is included in the totality of *οὐσία*, and every single being is an essence in the fundamental sense of being a something and not a nothing. Point (3) is the presence of the whole activity in a way proper to each particular. On this level the activity is received as a lot of different aspects (cf. Ch. 4 § IV).

That there is a common and a special presence does not mean that the activity itself is divided and distributed. The point here is that God is present as a whole to the creature and that the creature is present to God. The Maximian idea of participation is that God has introduced certain schemes which regulate the creature's relation to the activity. The creaturely presence is from the beginning defined by divine principles which delimit the receptive capacity (*ἐπιτηδειότης*) by which the creature emerges from non-being into being. These divine schemes or principles are the triadic scheme of *logoi*.

To illustrate this interpretation I will use an imperfect image. We place a magnet on the table with a sheet of paper over it. On the paper we spread iron filings. Everyone knows what will happen: the tiny bits of iron will immediately be arranged in a certain pattern. For the sake of the argument, let us make some distinctions which are not possible in the actual case: let us distinguish between (i) the magnet, (ii) the drawing-power of the magnet, (iii) the principle of the pattern and (iv) the pattern itself. The magnet is the separate entity which executes the drawing-power. The drawing-power by itself is always one and the same. The principle of the pattern makes the conditions on which the drawing-power is received in a medium. The pattern emerges in the way it does according to the principle of the pattern which delimits the effect of the drawing-power. Likewise, God, by Himself, transcends creatures. By the activity He is always present in the same way. The principles, i.e. the *logoi*, define the conditions on which the influence of the activity is received. According to this twofold divine operation (i.e. of the activity and the *logoi*) the created entities emerge into a certain modified presence.

The received *ἕξις* or *habitus*, mentioned in connection with the vocabulary of participation in section I above, is not to be understood in the way Balás, for instance, seems to take it (cf. Ch. 4 § 2). *Τὸ μετεχόμενον*, according to Balás, is either to be understood as the transcendent divine perfection or as this perfection as possessed by us. As far as I can understand, what is participated, as Balás sees it, is a created quality held by the creature. For Maximus, on the other hand, a received *ἕξις* could be nothing else than God Himself, viz. by the presence of His activity. According to St Maximus, created beings, in their essential otherness from God, are drawn into the presence of being and deification by the presence of the divine activity itself.

Concluding Remarks

In the preceding chapters I have attempted to penetrate the basic structure of St Maximus' thought. It has not been an easy task. What we have found is a Christian philosophy, which describes the system of the cosmos in accordance with basic Christian beliefs. Many of the ontological structures that I have discussed seem to fit together quite well. The doctrines of the *logoi*, the activities, the threefold structure of created beings, the threefold spiritual development, participation etc., all of this gives the impression that we are faced with a coherent world-view. But even though this philosophy seems to be coherent and consistent, the *mystery* of God's activity as Creator and Redeemer, as well as the *mystery* of God's final purpose for man and all created beings, emerge in every aspect of the Confessor's thought. The central focus of his thought is the drama of human life in a finite and transient world full of insecurity, a drama that gets its explanation in the light of God's economical activities.

By the end of this long treatment one of the questions that could be posed concerns the contemporary relevance of Maximus' thought. Does his philosophy belong to a past which long ago was superseded? I think not. But how could that be, since what we have found is a world-view which is tinted with philosophical realism and teleology—forms of thought which are suspect from a modern point of view? Seen from the angle of modern science the Maximian metaphysics would seem quite awkward and outdated. On the other hand, seen from the point of view of orthodox Christianity, modern science tends to misrepresent the world. It is not necessarily the case that modern science is simply false. Its major weakness, however, is that nature is interpreted from just one angle, i.e. its basic structure is seen as quantitative and material. It does not seem to me to be more reasonable to interpret the totality of Being in this way than to view it as comprising both a sensible and an intelligible reality. And if the totality of Being has both a sensible and an intelligible dimension, if the world is both matter and spirit, then to see it from just one angle means that the world-view could become more or less distorted. All of this requires a more comprehensive treatment and would exhaust the limits of the present investigation of Maximus.

However this may be, Maximus' kind of philosophy lives on in the liturgical experience of orthodoxy. His thought was not only then, several hundred years ago, congenial with the Christian experience of the meaning of Being, but is still so. The created world emerges, if not in its immediate actuality so at least in potentiality, as a community of beings. Creation is a unity-in-plurality, based on divine principles which are the means by which God seeks to accomplish the mystery of Christ. The Maximian philosophy has a great potential which could be utilized by the Church to address some urgent contemporary problems. I have especially two such modern issues in mind: (*a*) human rights and (*b*) the ecological issue. I shall end my treatment of Maximus' Christocentric cosmology with some comments on these two challenges.

(*a*) There should be no need to show why the human rights issue is important. It is, however, important to show how the Church might address it. From the point of view of orthodox Christianity as from the point of view of Maximus' thought, it seems strange to say that individual human beings have natural rights; we should at least have to address the subject from another angle. According to Maximus, there is a *logos* of the common human nature which all men have. We should remember that each *logos* is a representative of Christ, the Logos, so that there is a fundamental and unique relation between humanity as a whole and God. God created man out of His Goodness and philanthropy. Human beings are not just thrown into the world; they are there to move in accordance with nature towards the fulfilment of their being in God. Within historically existing humanity differentiations among human beings emerge that are unnatural and sinful to the degree that they prevent the actualization of the divine plan for human happiness. In the *Or. Dom.* Maximus comments on the phrase of St Paul that in Christ 'there is neither Greek nor

Jew, circumcision nor uncircumcision, barbarian, Scythian, bond nor free, but Christ is all, and in all'. Maximus talks of the tension of will that has pushed the one human nature to revolt against itself. This has introduced 'the unnatural law of mutual slaughter' among human beings. He goes further: ' "Neither slave nor freedman", that is, no division of the same nature by opposition of will, which makes dishonoured what is by nature of equal honour and has as an auxiliary the attitude of those who exercise a tyrannical sway over the dignity of the image.'[1]

There are two expressions of great interest here: (i) '[that] which makes dishonoured what is by nature of equal honour' (*ἄτιμον ποιουμένη τὸν κατὰ φύσιν ὁμότιμον*) and (ii) 'the dignity of the image' (*τὸ τῆς εἰκόνος ἀξίωμα*). The key terms are *τιμή*, 'honour', and *ἀξίωμα*, 'worth' or 'dignity'. Because the Logos by expressing Himself in their archetype (*logos*) has defined their common nature, human beings have by nature a certain honour and dignity. There is equality of honour and dignity because of the common basic principle of the nature. This does not imply that the focus is wholly on what is common, rather it implies that each human person is a unique instantiation of the nature that has the character of the divine image, and therefore is invested by God with this equality of honour and dignity.

The honour and dignity of human beings are guarded by God himself. In this world, the equality of honour and the dignity of the image should be secured in the Church, which is the body of Christ. God, according to Maximus, works for the unification and communion between all created beings, and this is what the Church does as well, being the image of God.[2]

The ontological structure of the world, the unity between all men established by God through the immanent taxonomic arrangement of the world, makes it a sin to discriminate human beings because of accidental social and ethnic relations. Rather, it is good to treat all men equally.[3]

As far as I can see this does not mean that the honour and dignity of man is given over to be preserved only in the relationship between

[1] Cf. St Paul: Gal. 3: 28; Col. 3: 11. *Or. Dom.*, CCSG 23: lines 478 ff.
[2] *Myst.* ch. 1, PG 91: 665c. [3] *De char.* 2.30, PG 90: 993b.

individual persons and in the community of the Church. From the point of view of Maximus' ontology, and from what he says in the quotations above, he seems to deliver a critique of injustice in social systems as well. From the point of view of the divine world-order it is not allowed to make social relations or social systems which in an unnatural way divide the common humanity and violate the honour and dignity of the human person as an image of God. Neither is it permitted to discriminate among persons because of their ethnic origin. From this point of view, it is natural to say that society is obliged to protect the human person from repression and violence, and to secure its possibilities to make free choices which do not violate the possibilities of other human beings.

(*b*) As we have seen in *Ambiguum* 41, Maximus points to the human being as a microcosm, created as the natural bond with the task of uniting all the levels of the created world with God. The ontological possibility of actualizing this task lies in the taxonomic system of beings. It is self-evident that given such a task and being made to exist within such possibilities, man was never allowed to suppress and exploit nature according to his own desires. Rather he should preserve, protect, and perfect his surroundings. The high value of the natural world is emphasized in Maximus' doctrine of *logoi* for genera, species, and individuals. The Logos-Christ not only expresses Himself in His *logos* for the human species and His *logoi* of human persons, He expresses Himself in the *logoi* of every created nature, of animals, insects, plants, minerals, etc., as well. By the simple fact of man's task and by the simple fact that the Logos is the foundation of the nature of all kinds and particulars, it is obvious that, according to God's will, the natural world should be treated with respect. It should not be made to suffer; it should not be exploited or violated. This, however, gives rise to some important questions. Man needs food and drink; he makes clothes and builds houses to protect him from bad weather. Is he allowed to feed on plants and animals and to use natural material (fur from animals, for instance) to make the things he need? The problem is that Maximus does not speculate over the human condition before the fall. Such speculation could have furnished us with a clue for understanding how he would consider problems of this kind. However, there are indications which seem to suggest that he accepted some ideas from St Gregory of

Nyssa concerning what happened to man as a consequence of the fall. Human bodily existence suffered a change. Because of this event man became corruptible and mortal. He began to propagate his race through sexuality, and to feed upon his surroundings. Corruptibility, mortality, and sexuality are nevertheless not sinful in themselves. They even have a providential aspect. Mortality, for instance, sets an end to sinful existence in the body, and sexuality makes it possible for man to leave behind someone of his own kind, thereby giving him comfort. To use our natural surroundings to fulfil basic needs is not in itself sinful, but here there is nevertheless a limit which should not be overstepped. Man is still a microcosm and his divinely given task still exists, even though it is to be fulfilled in Christ through the Church. Man therefore, is not allowed to exploit his environment unduly by letting his desires be the driving force in his activities.

Bibliography

Not all of the works referred to in the footnotes are given with the year that they were first published; rather I have given the year of the edition I have used. In the bibliography the original year of publication is given.

Works of St Maximus

Patrologia cursus completus, Series graeca, ed. J. -P. Migne, vols. 90–1 (Paris 1860.

Quaestiones ad Thalassium I, CCSG 7, 1980.

Quaestiones ad Thalassium II, CCSG 22, 1990.

Quaestiones et dubia, CCSG 10, 1982.

Opuscula exegetica duo: Expositio in psalmum LIX, Expositio orationis Dominicae, CCSG 23, 1991.

Ambigva ad Thomam, CCSG 48, 2002.

Translations

Berthold, Maximus the Confessor, *Selected Writings*, trans. and notes by G. C. Berthold, The Classics of Western Spirituality (London 1985).

Blowers, Paul M., and Robert L. Wilken, *On the Cosmic Mystery of Jesus Christ* (New York 2003).

Farrell, *The Disputation with Pyrrhus of our Father among the Saints, Maximus the Confessor*, trans. Joseph P. Farrell (South Canaan, Pa. 1990).

Larchet et al., Saint Maxime le Confesseur, *Ambigua*, introduction par J. -C. Larchet, avant-propos, traduction et notes par E. Ponsoye, commentaires par le Père D. Staniloae (Paris 1994).

Louth, A., *Maximus the Confessor* (London and New York 1996).

Palmer et al., *The Philocalia*, The Complete Text, trans. and ed. by G. E. H. Palmer, Ph. Sherrard, and K. Ware, vol. 2 (London and Boston 1981; repr. 1984).

von Ivánka, Maximos der Bekenner, *All-eins in Christus*, Auswahl, Übertragung, Einleitung von Endre von Ivánka (Einsiedeln 1961).

Other Ancient and Medieval Authors: Texts and Translations

For the works of Plato and Aristotle I have mostly used the editions in the Loeb Classical Library and the Oxford Classical Texts. For the works of Philo and Plotinus I have used the Loeb edition, and for the Neoplatonic commentators Porphyry, Ammonius, and Dexippus I have used the critical editions of the Berlin Academy (CAG), vol. 4, parts 1, 2, and 4. For Christian authors I have used the texts from Migne, *Patrologia graeca* (PG), whenever critical editions were not available. Other editions are mentioned below.

Alcinous, *The Handbook of Platonism*, ed. and trans. J. Dillon (Oxford 1995).

Alexander of Aphrodisias, *On Aristotle Metaphysics 1*, trans. W. E. Dooley, The Ancient Commentators on Aristotle (London 1989).

—— *On Aristotle Metaphysics 2 and 3*, trans. W. E. Dooley and A. Madigan (London 1992).

Ammonius, *On Aristotle Categories*, trans. S. Marc Cohen and G. B. Matthews, The Ancient Commentators on Aristotle (London 1991).

Aristotle, *Categories and De Interpretatione*, trans. with notes by J. L. Ackrill (Oxford 1974; first pub. 1963).

—— *The Complete Works of Aristotle*, ed. J. Barnes, 2 vols. (Princeton 1985).

—— *Metaphysics*, trans. H. G. Apostle (Bloomington and London 1975; first pub. 1966).

—— *Physics*, trans. H. G. Apostle (Bloomington and London 1980; first pub. 1969).

—— *Physics*, ed. Sir D. Ross (Oxford 1979; first pub. 1936).

Athanasius, *Contra Gentes and De Incarnatione*, ed. and trans. R. W. Thomson (Oxford 1971).

Basil, St, *Exegetic Homilies*, trans. Sister Agnes Clare Way, The Fathers of the Church (Washington 1981; first pub. 1963).

—— *On the Holy Spirit*, trans. D. Anderson, St Vladimir's Seminary Press (New York 1980).

—— *The Letters*, 4 vols., trans. R. J. Deferrari, Loeb Classical Library (London 1972; first pr. 1926).

Comnène, Anne, *Alexiad* (Paris 1967).

Dexippus, *On Aristotle Categories*, trans. J. Dillon, The Ancient Commentators on Aristotle (London 1990).

Diogenes Laertius, *Lives of Eminent Philosophers*, trans. R. C. Hicks, Loeb Classical Library (London 1980; first pub. 1925).

Eunomius, *The Extant Works*, text and trans. R. P. Vaggione (Oxford 1987).

Gregory of Nazianzus ['the Theologian'], *Poemata Arcana*, ed. C. Moreschini, D. A. Sykes, and L. Holford-Strevens (Oxford 1997).

Gregory of Nyssa, St, *Dogmatic Treatises, Etc.*, The Nicene and Post-Nicene Fathers, 2nd ser., vol. 5 (Peabody, Mass. 1995; first pub. 1893).

—— *On the Soul and the Resurrection*, trans. C. P. Roth, St Vladimir's Seminary Press (New York 1993).

—— *Opera, Contra Eunomium Libri*, ed. W. Jaeger, vols. 1 and 2 (Leiden 1960).

—— *La Vie de Moïse*, ed. J. Daniélou, Sources Chrétiennes (Paris 1968).

—— *The Life of Moses*, The Classics of Western Spirituality, trans. E. Ferguson and A. J. Malherbe (New York, Ramsey, Toronto 1978).

Gregory Palamas, *The One Hundred and Fifty Chapters*, a critical edn., trans., and study by R. E. Sinkewicz (Toronto 1988).

—— *The Triads*, The Classics of Western Spirituality, trans. N. Gendle (Paulist Press, NJ 1983).

—— *Défence des saints hésychastes*, vols. 1–2., introduction, texte critique, traduction et notes, Jean Meyendorff (Louvain 1959).

Ivstini philosophi et martyris, *Opera* (Wiesbaden 1969; Unveränderter Neudruck der 3. Auflage von 1876).

Justin Martyr, *Writings*, trans. T. B. Falls, The Fathers of the Church (Washington 1977; first pub. 1948).

Long, A. A. and Sedley, D. N., *The Hellenistic Philosophers*, vols. 1–2 (Cambridge 1992; first pub. 1987).

Origen, *Commentary on the Gospel according to John, books 1–10*, trans. R. E. Heine, The Fathers of the Church (Washington 1989).

—— *On First Principles*, trans. G. W. Butterworth (London 1936).

—— *Vier Bücher von den Prinzipien*, herausgegeben, übersetzt mit kritischen und erläuternden Anmerkungen versehen von H. Görgemanns und H. Karpp (Darmstadt 1976).

Philoponus, *Against Aristotle on the Eternity of the World*, trans. Chr. Wildberg, Ancient Commentators on Aristotle (London 1987).

—— *On Aristotle On the Soul 3. 1–8*, trans. W. Charlton, Ancient Commentators on Aristotle (London 2000).

Plotinus, *Ennead V. 1*, comm. and trans. M. Atkinson (Oxford 1983).

Porphyry the Phoenician, *Isagoge*, trans., introd., and notes, E. W. Warren (Toronto 1975).

Proclus, *The Elements of Theology*, trans., introd., and comm., E. R. Dodds, 2nd edn. (Oxford 1977; first pub. 1933).

Pseudo-Dionysius, *The Complete Works*, trans. C. Luibheid, The Classics of Western Spirituality (London 1987).

Stephani, *In liberum Aristotelis De interpretatione Commentarium*, ed. M. Hayduck, CAG vol. XVIII. iii (Berlin 1885).

Telfer, W. (ed.), *Cyril of Jerusalem and Nemesius of Emesa*, The Library of Christian Classics (London 1955).

Theophilus of Antioch, *Ad Autolycum*, ed. Grant, R. M. (Oxford 1970).

Other Works

Armstrong, A. H. (1979), 'The Background of the Doctrine "That the Intelligibles are not Outside of the Intellect" ', in *Plotinian and Christian Studies*, Variorum Reprints (London).

—— (ed.) (1980), *The Cambridge History of Later Greek and Early Medieval Philosophy* (Cambridge; first pub. 1967).

Balás, D. L. (1966), *ΜΕΤΟΥΣΙΑ ΘΕΟΥ, Man's Participation in God's Perfections according to Saint Gregory of Nyssa* (Rome).

Balthasar, Hans Urs von (1961), *Kosmische Liturgie, Das Weltbild Maximus' des Bekenners*. Zweite, völlig veränderte Auflage (Einsiedeln).

—— (1995), *Presence and Thought* (San Francisco; trans. from *Présence and pensée*, Paris 1988).

Bathrellos, D. (2004), *The Byzantine Christ: Person, Nature and Will in the Christology of St Maximus the Confessor* (Oxford).

Blowers, Paul M. (1991), *Exegesis and Spiritual Pedagogy in Maximus the Confessor* (Notre Dame, Ind.).

Boniz, H. (ed.) (1975), *Aristotelis opera vol. V, Index aristotelicus* (Berlin; repr. of 1870 edn.).

Carabine, Deirdre (1995), *The Unknown God* (Louvain).

Cherniss, H. (1934), University of California Publications in Classical Philology, 11 (1930–1933), 1–92.

Christou, Panayotis (1982), 'Maximos Confessor on the Infinity of Man', in F. Heinzer and Ch. Schönborn (eds.), *Maximus Confessor*, Actes du Symposium sur Maxime le Confesseur Fribourg, 2–5 septembre 1980 (Fribourg).

Clarke, Norris (1982), 'The Problem of the Reality and Multiplicity of Divine Ideas in Christian Neoplatonism', in Dominic J. O'Meara (ed.), *Neoplatonism and Christian Thought* (Albany, NY).

Cooper, A. G. (2005), *The Body in St Maximus the Confessor* (Oxford).

Copleston, F. (1962), *A History of Philosophy*, vol. 2, part 2 (New York).

Cornford, F. M. (1937), *Plato's Cosmology: the Timaeus of Plato translated with a running commentary* (London).

Dillon, John (1977), *The Middle Platonists* (London 1977).

—— (1985) review of *Studies in Platonism and Patristic Thought*, by John Wittaker, *Phoenix* 39(4), 420–2.

—— (1995), *Alcinous: The Handbook of Platonism* (Oxford).

Dörrie, H. (1959), *Porphyrios' 'Symmikta zetemata'* (Munich).

Duhem, Pierre (1985), *Medieval Cosmology* (Chicago and London).

Edwards, M. J. (2002), *Origen against Plato* (Aldershot).

Emilsson, E. K. (1994), 'Platonic Soul-Body Dualism in the Early Centuries of the Empire to Plotinus', in W. Haase and H. Temporini (eds.), *Aufstieg und Niedergang der römischen Welt*, II. 36. 7, 5332.

—— (1996), 'Cognition and its Object', in L. P. Gerson (ed.), *The Cambridge Companion to Plotinus* (Cambridge).

—— (1999), 'Remarks on the Relation between the One and Intellect in Plotinus', in John Cleary (ed.), *Traditions of Platonism* (New York).

Farrell, J. (1989), *Free Choice in St Maximus The Confessor* (South Canaan, Pa.: St Tikhon's Seminary Press).

Frede, M. (1987), *Essays in Ancient Philosophy* (Oxford).

Friedrich, Gerhard (ed.) (1983), *Theological Dictionary of the New Testament*, vol. 7, trans. and ed. by G. W. Bromiley, Grand Rapids, Michigan (repr. of 1971 edn.).

Gersh, Stephen (1978), *From Iamblichus to Eriugena, An Investigation of the Prehistory and Evolution of the Pseudo-Dionysian Tradition* (Leiden).

Gilson, E. (1980), *History of Christian Philosophy in the Middle Ages* (London; first pub. 1955).

Grillmeier, A. (1975), *Christ in Christian Tradition*, vol. 1 (London and Oxford; first pub. in Eng. 1965).

—— (1995) in collaboration with Hainthaler, T., *Christ in Christian Tradition*, vol. 2, part 2 (London; first pub. in Ger. 1989).

Guthrie, W. K. C. (1986), *A History of Greek Philosophy, Vol. V: The later Plato and the Academy* (Cambridge; first pub. 1978).

Habra, G. (1957–8), 'The Sources of the Doctrine of Gregory Palamas on the Divine Energies', *Eastern Churches Quarterly*, 12: 244–52, 294–303, 338–47.

Joannou, P. (1956), *Christliche Metaphysik in Byzanz* (Ettal).

Kattan, A. E. (2003), *Verleiblichung und Synergie* (Leiden).

Kazhdan, A. P. (ed.) (1991), *The Oxford Dictionary of Byzantium*, vol. 1–3 (Oxford).

Lampe, G. W. H. (ed.) (1989), *A Patristic Greek Lexicon* (Oxford; first pub. 1961).

Larchet, Jean-Claude (1996), *La Divinisation de l'homme selon saint Maxime le Confesseur* (Paris).

—— (1998), *Maxime le Confesseur, médiateur entre l'Orient et l'Occident* (Paris).

Lloyd, A. C. (1981), *Form and Universal in Aristotle*, ARCA, Classical and Medieval Texts, Papers and Monographs 4.

—— (1990), *The Anatomy of Neoplatonism* (Oxford).

Lossky, V. (1973), *The Mystical Theology of the Eastern Church* (Cambridge and London; this trans. first pub. 1957).

—— (1983), 'Tradition and traditions', in L. Ouspensky and V. Lossky, *The Meaning of Icons* (St Vladimir's Seminary Press; first pub. 1952), 9–22.

Louth, A. (1998), 'Recent Research on St Maximus the Confessor: A Survey', *St Vladimir's Theological Quarterly* 42(1), 67–84.

Lovejoy, A. O. (1978), *The Great Chain of Being* (Cambridge, Mass., and London; first pub. 1936).

Meyendorff, J. (1974), *A Study of Gregory Palamas* (St Vladimir's Seminary Press; first pub. in Eng. 1964).

—— (1975), *Christ in Eastern Christian Thought* (St Vladimir's Seminary Press).

—— (1987*a*), *Byzantine Theology* (New York; 1st edn. 1974).

—— (1987*b*), 'Christ's Humanity: The Paschal Mystery', *St Vladimir's Theological Quarterly* 31(1), 5–40.

Mueller-Jordan (2005), *Typologie spatio-temporelle de l'Ecclesia Byzantine, La Mystagogie de Maxime le Confesseur dans la culture philosophique de l'Antiquité tardive* (Leiden).

Nichols, Aidan (1993), *Byzantine Gospel. Maximus the Confessor in Modern Scholarship* (Edinburgh).

O'Meara, Dominic (1980), 'The Problem of Omnipresence in Plotinus *Ennead* VI, 4–5: A Reply', *Dionysius* 4, 61–74.

Owen, G. E. L. (1979), 'Logic and Metaphysics in Some Earlier Works of Aristotle', in J. Barnes, M. Schofield, and R. Sorabji (eds.), *Articles on Aristotle*, vol. 3: *Metaphysics* (London), 13 ff.

Pelikan, Jaroslav (1977), *The Spirit of Eastern Christendom* (Chicago and London).

—— (1993), *Christianity and Classical Culture* (New Haven and London).

Perl, E. D. (1991), *Methexis: Creation, Incarnation, Deification in Saint Maximus the Confessor*, Yale University (dissertation).

Peters, F. E. (1967), *Greek Philosophical Terms* (New York and London).

Prestige, G. L. (1981), *God in Patristic Thought* (London; first pub. 1936).

Quasten, J. (1994), *Patrology*, vol. 3 (Westminster, Md.; first pub. 1950).

Riou, A. (1973), *Le monde et l'église* (Paris).

Rist, J. M. (1967), *Plotinus: The Road to Reality* (Cambridge).

—— (1996), 'Plotinus and Christian philosophy', in L. P. Gerson (ed.), *The Cambridge Companion to Plotinus* (Cambridge).

Ritter, J. (1972), *Historisches Wörterbuch der Philosophie*, vol. 2 (Darmstadt).

Roueché, M. (1974), 'Byzantine Philosophical Texts of the Seventh Century', *Jahrbuch der österreichischen Byzantinistik*, vol. 23 (Vienna), 61–76.

—— (1980), 'A Middle Byzantine Handbook of Logical Terminology', *Jahrbuch der österreichischen Byzantinistik*, vol. 29 (Vienna), 71–98.

Sambursky, S. (1962), *The Physical World of Late Antiquity* (London).

Sellers, R. V. (1953), *The Council of Chalcedon* (London).

Sherwood, P. (1950), 'Notes on Maximus the Confessor', *The American Benedictine Review* 1(1), 347–56.

—— (1952), *An Annotated Date-list of the works of Maximus the Confessor* (Rome).

—— (1955), *The Earlier Ambigua of Saint Maximus the Confessor* (Rome).

—— (1964), 'Survey of Recent Work on St Maximus the Confessor', *Traditio* 20, 428–37.

Siddals, R. M. (1987), 'Logic and Christology in Cyril of Alexandria', *Journal of Theological Studies*, NS 38, 341–67.

Sleeman, J. H. and Pollet, G. (eds.) (1980), *Lexicon Plotinianum* (Leiden).

Sorabji, Richard (1983), *Time, Creation and the Continuum* (London).

—— (ed.) (1987), *Philoponus and the Rejection of Aristotelian Science* (London).

—— (1988), *Matter, Space and Motion* (London).

—— (ed.) (1990*a*), *Aristotle transformed* (London).

—— (1990*b*), 'The Ancient Commentators on Aristotle', in Sorabji (1990*a*), 1–30.

Spearritt, P. (1968), *A Philosophical Enquiry into Dionysian Mysticism* (Fribourg).

Stead, Ch. (1996), *Philosophy in Christian Antiquity* (Cambridge; first pub. 1994).

Tanner, N. P. (ed.) (1990), *Decrees of the Ecumenical Councils*, 2 vols. (London and Washington, DC).

Thunberg, L. (1985), *Man and the Cosmos* (New York).

—— (1995), *Microcosm and Mediator* (Chicago and La Salle; 1st edn. Lund).

—— (1999), *Människan och kosmos* (Artos bokförlag).

Tollefsen, T. (1992), 'Aristoteles om ousia', in *Opuscula—Examen philosophicums småskriftserie* nr. 1 (Oslo).

Törönen, M. (2007), *Union and Distinction in the Thought of St Maximus the Confessor* (Oxford).

Turcescu, L. (1997), 'The Concept of Divine Persons in Gregory of Nyssa's *To His Brother Peter, on the Difference Between* Ousia *and* Hypostasis', *The Greek Orthodox Theological Review* 42(1–2), 63–82.

Verrycken, K. (1990), 'The Metaphysics of Ammonius son of Hermeias', in Sorabji (1990*a*), 199–231.

Wesche, K. P. (1987), 'The Christology of Leontius of Jerusalem, Monophysite or Chalcedonian?' *St Vladimir's Theological Quarterly* 31(1), 65–95.

Whittaker, John (1984), *Studies in Platonism and Patristic Thought*, Variorum Reprints (London).

Wilson, N. G. (1983), *Scholars of Byzantium* (London).

Wippel, J. F. (1993), *Thomas Aquinas on the Divine Ideas* (Toronto).

Witt, Charlotte (1989), *Substance and Essence in Aristotle* (Ithaca, NY and London).

Wolfson, H. A. (1976), *The Philosophy of the Church Fathers* (Cambridge, Mass. and London; first pub. 1956).

Zizioulas, John D. (1985), *Being as Communion* (New York).

Index

CPSIA information can be obtained at www.ICGtesting.com
Printed in the USA
LVOW11*0409040916

503140LV00004B/21/P

9 780199 237142